Lazzaro Spallanzani

Dissertations Relative to the Natural History of Animals and Vegetables

Vol. 1

Lazzaro Spallanzani

Dissertations Relative to the Natural History of Animals and Vegetables
Vol. 1

ISBN/EAN: 9783337376000

Printed in Europe, USA, Canada, Australia, Japan

Cover: Foto ©berggeist007 / pixelio.de

More available books at **www.hansebooks.com**

DISSERTATIONS

RELATIVE TO THE

NATURAL HISTORY

OF

ANIMALS AND VEGETABLES.

TRANSLATED FROM THE ITALIAN OF THE

ABBÉ SPALLANZANI,

Royal Profeſſor of Natural Hiſtory in the Univerſity of PAVIA, Superintendent of the PUBLIC MUSEUM, and FELLOW of various learned SOCIETIES.

In TWO VOLUMES.

A NEW EDITION CORRECTED AND ENLARGED.

VOL I.

LONDON:

PRINTED FOR J. MURRAY, NO. 32, FLEET-STREET.

M DCC LXXXIX.

INTRODUCTION.

IN the courfe of my public demonftrations in the year 1777, I repeated in the prefence of my hearers thofe celebrated experiments of the Academy of Cimento, which fhew that the ftomachs of fowls and ducks exert fo aftonifhing a force as to reduce hollow globules of glafs to powder in the fpace of a few hours. Finding them perfectly exact, I conceived the defign of extending them to fome other individuals of that clafs of birds which have been termed birds with *mufcular ftomachs* or *gizzards*. Such were the firft lines of an undertaking, of which till that time I had never entertained the fmalleft idea, and which afterwards became more and more extenfive, as my curiofity concerning fo fine and ufeful a fubject as the important function of Digeftion increafed. Hence from animals with mufcular ftomachs, I was induced to proceed to thofe with intermediate, and from thefe again to animals with membranous ftomachs.

machs (*a*). Thus I enjoyed the pleasure of extending my researches to the principal classes of animals, not neglecting Man, the noblest and most interesting of all. But these physiological researches laid me under the necessity of examining the most celebrated systems concerning Digestion, and of enquiring whether it is effected by trituration, by a solvent, by fermentation, or by an incipient putrefaction: or whether, according to the opinion of the great Boerhaave, it rather depends upon all these causes operating in conjunction. Thus I was obliged to enter anew upon a question of very ancient date, and though discussed at great length by many physiologists, yet not in my opinion sufficiently elucidated; since most writers have chosen to follow the delusive invitation of theory and hypothesis, rather than the unerring direction of decisive experiments. The impartial and judicious reader, when he shall have perused the present essay, will be able to determine, whether what I assert, be true or false.

(*a*) The I, LVIII, and CIV paragraphs will explain what is meant by birds with muscular, intermediate, and membranous stomachs.

DISSERTATION I.

CONCERNING

DIGESTION.

ON THE DIGESTION OF ANIMALS WITH MUSCULAR STOMACHS, COMMON FOWLS, TURKEYS, DUCKS, GEESE, DOVES, PIGEONS.

I. THOUGH there perhaps exists no animal, of which the stomach is not furnished with muscles, yet there is a singular class, justly denominated by several naturalists *animals with muscular stomachs*, since that viscus is provided with remarkably large and powerful muscles. To this class belong fowls, ducks, pigeons, geese, partridges, &c. So great is the strength of these muscles, that many have imagined that they produce digestion by acting violently upon the contents of the stomach, and breaking down and re-

ducing

ducing them to a pultaceous mafs, in no refpect differing from imperfect chyle. This notion was afterwards applied to other animals, nor was man himfelf exempted; and it has been pretended, that digeftion is univerfally owing to the alternate action of the mufcles of the ftomach, or, as it has been termed, to *trituration*.

11. Now, to confine ourfelves to animals with mufcular ftomachs, there was little difficulty in devifing the means of determining whether the comminution and folution of food is effected by the gaftric mufcles. Such means have been contrived and fuccefsfully put in practice by Reaumur. " Let feveral animals refembling each other in ftructure," fays that great naturalift in his two excellent memoirs on this fubject, of which I fhall make frequent ufe in the fequel, " be made to fwallow metallic tubes open at both ends, and filled with fome of their natural food, as grains of the *Cerealia* when gallinaceous fowls are the fubjects of our experiments. Should thefe grains, after they have remained a certain time in the ftomach, be broken down and decompofed, we muft affume a diffolving liquor as the caufe of this phænomenon, fince the fides of the metallic tubes muft have been an infuperable obftacle to the exertions of the gaftric

gaftric mufcles upon the contents; but if they fhould be retrieved in a found and entire ftate, it muft be acknowledged, that in thefe animals digeftion does not depend on a folvent, but on the action of mufcles." And fuch was the plan adopted by this fagacious philofopher. He enclofed fome barley in metallic tubes open at each end and forced them down the throats of common fowls, turkeys and ducks. Upon killing the animals fome hours afterwards and taking the tubes out of the ftomach, the grains were found quite entire; whence he inferred, that in birds of the gallinaceous-clafs the food is not broken down by a folvent, but by ftrong mufcular action.

III. This experiment feems indeed highly favourable to the doctrine of trituration; yet I think it would have been much more conclufive, if the fame refult had been obtained from other individuals of this clafs, and if befides barley, other grains upon which they naturally feed, fuch as wheat, maize, rye, chickpeafe, &c. had been employed. I therefore refolved to put each of thefe feeds to the teft of experiment in the following manner. I procured fome tin tubes eight lines in length and four in diameter, and inclofed in each a quantity of feeds greater or lefs according as they

they were themselves smaller or larger. The ends of the tube were left open, but iron wires were made to pass before them, so as to cross each other, and form a kind of lattice-work. Common fowls were the first subjects of my experiments: I forced some of the tubes into the stomach, conducting them with my forefinger and thumb through the œsophagus, till I was certain they were in the cavity of that viscus. When this operation is properly performed, neither fowls nor other animals sustain any injury. In twenty-four hours the tubes were taken out, and the contents upon examination appeared to be unaltered: even the colour and taste were unchanged, if we except a slight bitter flavour which they had acquired. They had imbibed some moisture, and were a little swoln. The same seeds inclosed also in tubes, and left in the stomach two and even three days, underwent no greater change.

IV. Several times, immediately after having forced the tubes full of grains into the stomach, I introduced some of the same grains loose. The latter were broken down in a few hours, but the former remained entire.

V. The food taken spontaneously by these birds does not pass immediately into the stomach, but stops for some time in the crop,
where

where it is macerated, and becomes softer. Is such a previous maceration necessary before it can be dissolved within the tubes? This circumstance seemed to deserve attention. I therefore repeated the foregoing experiments with seeds taken from the crop of a fowl, after they had undergone a complete maceration. Notwithstanding this preparation, they underwent no change within the tubes.

VI. From these results it was easy to predict, that no new appearance would occur, if the skin should be taken off, as it really happened. It is proper to add, that other grains treated in the same manner were no more dissolved than those before-mentioned.

VII. The mode which I had hitherto practised of using tubes open at both ends, at which the gastric fluid was certainly at liberty to enter, was that of Reaumur. But this fluid having no other access, cannot exert its action on the inclosed grains so powerfully as when they are loose in the stomach, as Reaumur ingenuously confesses. To obviate this inconvenience in some measure, I had the sides of the tubes perforated with a great number of holes. I had moreover recourse to another expedient. I employed hollow globules of brass half an inch in diameter, and pierced like a sieve, which I could open and shut

shut at pleasure by means of a screw worked upon the edge of the two hemispheres, into which each globule was divisible. With these new tubes and spherules I repeated the preceding experiments, not only upon common fowls, but upon ducks, turkeys, geese, doves, and pigeons: and as a larger quantity of liquor could now find its way to the inclosed substances, they were more thoroughly soaked, and had acquired a bitterer taste (III); but I could never perceive the slightest token of solution, though they continued a long time in the stomach.

VIII. These facts afford an irrefragable proof, that the trituration of seeds in the stomach of granivorous birds, is solely owing to strong pressure and repeated and violent percussions: effects produced by the powerful muscles with which that organ is furnished.

IX. The contents of the stomach are so violently agitated as to be driven in at the open ends and through the holes of the tubes and spherules, which occasions some confusion. Hence I have frequently found it of service to introduce these receivers when the stomach was empty, and to keep the animal fasting during the whole time of the experiment.

x. The

x. The violent action of the fides upon the contents of the ftomach renders another precaution highly neceffary. The thicknefs of the tubes and fpherules fhould be confiderable, otherwife the obferver, when he takes them out of the ftomach, will find them broken, cruſhed, or diftorted in a moft fingular manner, if they have been long retained. Reaumur mentions feveral accidents of this fort (*a*); and I have feen inftances without number of fuch contufions, one of which I cannot forbear relating here. Having found that the tin tubes which I ufed for common fowls were incapable of refifting the force of the ftomach of turkeys, and not happening at that time to be provided with any tin foil of greater thicknefs, I tried to ftrengthen them by foldering to the ends two circular plates of the fame metal, perforated only with a few holes for the admiffion of the gaftric fluid. But this contrivance was ineffectual; for after the tubes had been twenty hours in the ftomach of a turkey, the circular plates were driven in, and fome of the tubes were broken, fome compreffed, and fome diftorted in the moft irregular manner.

(*a*) In the Memoir quoted above.

XI. I then tried the following means of preventing this inconvenience. Having perforated the circular laminæ in the center, I paſſed an iron wire through the holes, and bound it tight round the outſide of the tubes and twiſted the two ends together. And now though the ſoldering ſhould be deſtroyed, yet this contrivance would prevent the circular laminæ from receding from the ends of the tube, unleſs the wire which paſſed through them ſhould be broken. I prepared four tubes in this manner, and gave them to a turkey ſix months old. After they had remained a whole day in the ſtomach, I killed the animal; and was exceedingly ſurprized to find that the tubes, in ſpite of my expedient, were very much damaged. All the iron wires were broken, two where they were twiſted, and the two others at their entrance into the tubes: the laminæ, ſo far from remaining ſoldered to the tubes, were found amongſt the food; they were not flat as at firſt, but ſome were bent ſo as to form an angle, ſome curved, and in others, one part was preſſed cloſe to the other. The tubes had ſuſtained equal injury; two of them were flattened as if they had been ſtruck by an hammer, the third was moulded into the ſhape of a gutter, the ſoldering of the fourth was deſtroyed, and it was made as flat as a wafer. XII. Theſe

xii. These phænomena will not so much surprize those who have learned from Redi (*a*) and Magalotti (*b*), how ducks, fowls, and pigeons reduce to powder hollow globules of glass in a very short space of time, and even solid ones in a few weeks. I have already observed, that I repeated these experiments with the greatest success (*c*). Some spherules of glass blown by the lamp, and so thick that they would seldom break when thrown upon the ground, were commonly reduced to small fragments, after remaining three hours in the stomach of a hen or a capon; the fragments were not sharp as when they are broken by the efforts of the hand, but as obtuse as if their edges and points had been abraded by a grinding-stone. The longer the spherules continued in the stomach, the more minutely were they triturated; so that in a few hours they were reduced to a mass of particles, not larger than grains of sand. Moreover the rapidity of this process appears in some measure proportional to the size of the animal. A wood-pigeon generally breaks them less speedily than a chicken, a chicken than a

(*a*) Esperienze intorno a cose naturali.
(*b*) Saggio di naturali esperienze.
(*c*) In the Introduction.

capon,

capon, but a goose the soonest of all. The reason is plain, since the larger species have thicker and more powerful stomachs.

XIII. From these and other facts which I shall adduce hereafter, we may collect how much the celebrated Pozzi, formerly Professor at Bologna, was mistaken, when he considered the observations (*a*) of the Florentine Academicians, and of Redi on the power of certain animals to reduce globules of glass to pieces as false, because he failed in his attempts to repeat their experiments on pigeons. Let me here be allowed to remark that it is the custom of certain dabblers in philosophy to deny facts, however particularly described, and though related by persons of the highest authority, merely because their own endeavours fail of success. But they do not reflect, that this is acting in direct opposition to the principles of sound logic, by which we are taught that a thousand negative facts cannot destroy a single positive fact, since it is so very easy to omit some one or other of the many circumstances requisite to the success of an experiment. The Bolognian Physician has fallen into this error; instead of so rashly infering from his own observation the falsity of the

(*a*) In his short anatomical essay printed at Bologna by Lælius a Vulpe.

contrary

contrary event, he ought to have multiplied and varied his experiments; and if he had done this with proper precautions, he would have confirmed, inftead of contradicting the relation of the Florentine philofophers. We muft fuppofe, that the ftomachs of his pigeons were too weak and flaccid to abrade and break fubftances of fuch hardnefs as glafs, from their being either in an unhealthy ftate, or too young; for in thefe cafes they are by no means capable of producing fuch effects, as I have found from actual experience.

XIV. My celebrated countryman Vallifneri, in his judicious *anatomy of the oftrich*, fuppofes that the hardeft fubftances, fuch as ftones, wood, glafs, and even iron itfelf, are reduced to pieces in the ftomach of this enormous bird by a folvent; he alfo inclines to think, that glafs is attacked and broken by fuch a liquor, which he imagines to exift in the ftomachs of fowls, without the concurrence of mufcular action. But the hypothefis of Vallifneri is evidently groundlefs; for feeds, as we have feen above, remain unaltered whenever they are defended by tubes. And when pigeons, fowls, or turkeys, are forced to fwallow feveral balls of glafs at once, fome inclofed in tubes, and others loofe, the latter are reduced to fmall fragments as ufual,

while

while the former remain entire. That the gaftric mufcles are the fole caufe of this effect, will appear ftill more evidently from facts to be related in the fequel (xv).

xv. Before I proceed farther in the recital of experiments immediately relating to digeftion, it may be proper to mention fome other phænomena analogous to thofe juft defcribed. They may help to convey more diftinct notions concerning this function in animals furnifhed with gizzards; the fmooth and blunt fubftances hitherto employed, could not injure the ftomach. It was therefore an object of curiofity to enquire what would happen when fharp bodies were introduced. It is well known how readily broken glafs will lacerate flefh. I therefore gave a cock feveral fragments of a broken pane, each about the fize of a pea; they were wrapped up in paper, to prevent the œfophagus from being torn as they paffed through it. I was well affured that this cover would be immediately deftroyed on its entrance into the ftomach, and leave the glafs at liberty to act with its points and edges. The animal was killed in twenty-four hours, and the glafs was found in the ftomach; but on this, as well as former occafions, the angles were fo far obliterated, that upon putting fome of the fragments on

the

the palm of one hand, and rubbing them forcibly on the back of the other, I did not receive the leaft hurt. Upon weighing the glafs, it appeared to have loft twenty-four grains; nor was it difficult to difcover what was become of the miffing particles, for the fides of the ftomach, when viewed attentively glittered with innumerable vitreous points. On the contrary, fome broken bits of glafs, that were inclofed in two tubes, of which one was given to a hen, and the other to a turkey, and left twenty-four hours in the ftomach, were not at all abraded at their points or edges.

xvi. Similar pieces of glafs, that remained two days in the ftomach of a wood-pigeon, gave me an opportunity of obferving other remarkable fractures and abrafions. The mention of this bird brings to my mind a fact relating to the prefent fubject. I gave a wood-pigeon an unpolifhed twelve-fided garnet, of the fize of a nut, with the intention of infpecting the ftomach a few hours afterwards; the bird was confined in a cage, but made its efcape by fome accident, and mixing among a number of others kept in another place, I was not able to diftinguifh it; fo that it did not fall into my hands for a month. The garnet, which had remained all this time in

the

the ſtomach, filled almoſt its whole capacity; a circumſtance which a little ſurprized me, ſince it had taken its food, and been nouriſhed very well. But I was ſtill more ſurprized at finding the angles of this hard ſtone blunted in ſome places.

XVII. But the reader will ſurely be eager to learn what injury the ſtomach received from the violent agitation thoſe ſharp bodies muſt have undergone during the abraſion of their moſt pointed parts. To ſatisfy my own curioſity, as well as that of others, I opened the cock and the two wood-pigeons (XV, XVI), and examined the internal coat of the ſtomach with the cloſeſt attention, after having waſhed away the contents. I moreover diſſected it away from the nervous coat; this was eaſily effected: and I could now examine it to greater advantage, but notwithſtanding all my pains, found it perfectly entire. No laceration, no diviſion, not the ſmalleſt jagged appearance; it was in every reſpect like ſtomachs that had not afforded reception to any unuſual ſubſtance. Only the coat of that ſtomach which had retained the large garnet for a month, was about three times as thick as it commonly is.

XVIII. Finding that theſe fowls ſuſtained theſe experiments unhurt, I ſubjected them to two others far more dangerous. Twelve ſtrong needles were firmly fixed in a ball of lead, in ſuch a manner that the points projected about a quarter of an inch from the ſurface. Thus armed, it was covered with a caſe of paper, and forced down the throat of a turkey. The bird retained it for a day and half without ſhewing the leaſt ſymptom of uneaſineſs. Why the ſtomach ſhould have received no injury from ſo horrid an inſtrument, I cannot explain: but the points of the twelve needles were broken off cloſe to the ſurface of the ball, except two or three of which the ſtumps projected a little higher. The ball had not loſt its general ſhape, but was marked with ſeveral indentations, that certainly were not upon it at firſt. Two of the points of the needles were found among the food, the other ten I could not diſcover either in the ſtomach or the long tract of the inteſtines; and therefore concluded that they had paſſed out at the vent.

xix. The ſecond experiment, ſtill more cruel, conſiſted in fixing twelve ſmall lancets, with very ſharp points and edges, in a ſimilar ball of lead. They were ſuch as I uſe for the diſſection of ſmall animals. The ball was given to a turkey cock, and left eighteen hours in the ſtomach; at the expiration

ration of which time it was opened, but nothing appeared but the naked ball, the twelve lancets having been broken to pieces; I discovered three in the large inteſtines, pointleſs and mixed with the excrements; the other nine were miſſing, and had probably been voided at the vent. The ſtomach was as found and entire as that which had received the needles.

xx. Of two capons, one ſuſtained the experiment with the needles, and the other that with the lancets, equally well. My next wiſh was to know how much time had elapſed before thoſe ſubſtances begin to be acted upon. By repeated experiments on turkeys that were killed after intervals ſucceſſively ſhorter, I found that theſe ſharp bodies begin to be broken and loſe their ſhape in about two hours. This at leaſt happened in two individuals of that ſpecies: in one four of the lancets, and in the other, three of the needles were broken within that ſpace; the reſt were blunted, but continued firm in the balls.

xxi. Let it not however be ſuppoſed, that the ſtomach in this claſs of birds is altogether invulnerable. In pullets it certainly is ſometimes very much injured. I obliged two pullets to ſwallow ſome pins without heads. One was killed in eight, and the other in
thirty-

thirty-two hours. The former had not at all suffered, but two pins were stuck in the stomach of the latter. These stomachs, as well as those of many other animals, are full of furrows, in one of which the two pins were fixed almost perpendicularly, one to the depth of a line, and the other to that of three lines: they were opposite to the most muscular part of the organ. Some force was required to extract them; at the puncture appeared a little clotted blood, with an evident livid colour around.

XXII. But whatever conclusion we are to draw from this last fact, it is certain, that the stomach of such birds is in general not subject to any injury from the introduction, residence, and trituration of these and the like substances, as I have learned from a vast variety of experiments. But how is it possible, some will enquire, that the gastric muscles can contuse, triturate, and even sometimes reduce to impalpable powder (as when glass is employed, XII, XIV, XV, XVI) these pointed bodies without injury to themselves? If the muscles act with so much force, must not the substances necessarily re-act upon the muscles? And will not this re-action cause the laceration of the internal coat of the stomach, which, though it is indeed very firm and compact, cannot surely sustain such violent shocks with impunity? XXIII. This

XXIII. This objection was immediately started, upon the discovery of the wonderful force with which digestion in poultry is effected, and an attempt was made to remove it in the following ingenious manner. It had been long known, that fowls, and other birds of the same class, have always a smaller or larger supply of little pebbles in their stomachs. It was therefore conceived, that these pebbles serve as a shield to the muscles. Hence it follows, that the comminution of bodies forced into the stomach is the immediate effect of the pebbles, and only the mediate effect of muscular action. Accordingly, the Academicians of Cimento have observed, that those ducks and fowls that have most stones in their stomachs, soonest reduce spherules of glass to powder. Redi thinks, that the stones perform the office of teeth (*a*); and Reaumur supposes them necessary to digestion (*b*).

XXIV. In the course of my numerous observations I can safely assert, that I never opened the stomach of a pigeon, turtle-dove, dove, partridge, fowl, turkey, goose, &c. without finding some small stones in it. I have also found what is remarked by Reaumur, that the size of the stones is apparently proportional to the size of the bird. They are generally of a roundish shape, whether

(*a*) L. c. (*b*) Mem. cit.

they

they acquire it from friction within the cavity of the ftomach, or have it before they are fwallowed. They are commonly bits of quartz, fometimes mixed with fragments of calcareous ftones. In the ftomach of a turkey hen I have counted above 200, and above 1000 in that of a goofe. Their exiftence is therefore indubitable. But is it equally certain, that they are the immediate inftruments of trituration? He who is unprejudiced in favour of any theory muft immediately perceive, that this is a mere hypothefis, convenient indeed and plaufible, but requiring to be confirmed by experiment.

xxv. To this teft I have endeavoured to bring it, and would willingly hope that I have decided the queftion. According to the obfervation of the Academicians, thofe birds that have moft ftones in the ftomach, fooneft triturate hard fubftances. Nothing was more eafy than to repeat the experiment. This I did upon ducks and fowls, the two fpecies mentioned by thofe learned writers, fometimes obliging them to fwallow globules of glafs, fometimes thin tubes of tin, and at others feeds defended by a ftronger cover, fuch as nuts of a moderate fize. It was neceffary that all circumftances fhould be alike, that the birds fhould be of the fame fpecies and age, and of equal vigour, &c. Not to weary the reader with too minute a detail, I shall only mention the results. In a hen and two ducks, not abundantly supplied with pebbles, the injury sustained by the substances was not so great as in three other like fowls more amply provided with them. But in four hens the effect produced was exactly the same, as far as I could judge, though the stomachs of three were less copiously furnished than that of the fourth.

xxvi. Having collected a large quantity of stones from the diffection of many gizzards, I thought they might be useful in the present inquiry; I therefore gave a certain number to some fowls and ducks, while others were left with those which they had swallowed spontaneously. The former, according to the observation of the Florentine Academicians, ought to have broken down hard substances sooner than the latter. And so indeed it sometimes happened, but at others the event was different. Wherefore not being able to ascertain the object of my enquiry by these experiments, I had recourse to other means of solving the problem.

xxvii. The most decisive mode of determining the use of stones in digestion, evidently was to take them away altogether, either

either by expelling thofe already fwallowed, or by preventing the admiffion of any at all. To evacuate thofe already accumulated, it was neceffary to confine the birds in cages where they could not find frefh ones, and it might be hoped, that the old ones would be gradually voided with the excrements. Accordingly, feveral fowls, turkeys, pigeons, and ducks were confined feparately, and that all fufpicion of their picking up pebbles might be removed, the cages were raifed to fuch an height that they could not reach the floor with their beaks. The bottoms were made of ofiers placed at a diftance from each other, in order that if the ftones fhould pafs out with the excrement they might not remain in the cage, and be fwallowed again, but fall to the ground. I fed them myfelf the whole time, taking care that the food, confifting of corn, vetches, and maize, fhould be free from all foreign matter, fo that I was certain not a fingle grain of fand or the fmalleft ftone was fwallowed by them.

XXVIII. In the courfe of a few days I perceived fome ftones among the excrement, and they continued to be voided during the whole time of confinement. Two days before the end of the month, when they were to be killed, I forced fome to fwallow tubes of tin, others glafs globules, others balls of lead, fome naked and fome armed with needles and lancets (XVIII, XIX, XX.) I likewife gave them fome grains of wheat, but did not allow them to undergo the natural procefs of maceration in the crop. On the 30th day every ftomach was carefully examined, and though I did not completely attain the end in view, yet I gained confiderable information on the fubject. Not a fingle ftomach indeed was free from ftones, but they were few in number, in fome inftances not amounting to above four or five, and thofe very fmall. The contufions, however, on the tin tubes, the indentations on the naked balls, the fracture of the needles and lancets, the trituration of the grain had alike taken place in every ftomach; nor did it appear, that the diminution of the quantity of ftones at all contributed to diminifh the alteration of the feveral fubftances, or to occafion any injury to the organ that contained them. And left it fhould be objected, that thefe hard bodies themfelves performed the office of pebbles by rubbing violently againft each other in confequence of the action of the gaftric mufcles (an objection manifeftly trivial) I had taken care that each bird fhould not have more than one tin tube, or one glafs globule, &c.

&c. Thefe folitary fubftances were juft as much bruifed or broken as when many were put into one ftomach; and that vifcus remained as free from injury.

xxix. Though thefe facts abundantly prove, that trituration does not depend on the ftones fwallowed by the birds in queftion, but upon the ftrength and action of the gaftric mufcles, I yet wifhed, by obferving what happens in ftomachs that have not received any ftones, for proofs ftill more decifive. The judicious reader will perceive at once, that to accomplifh my purpofe, it was neceffary to procure young neftlings that had never been abroad in queft of food. Accordingly, I procured fome wood-pigeons, yet unfledged, were brought me; but I was difappointed in my expectations, for even their tender ftomachs were not free from pebbles, which doubtlefs were mixed with the food carried to them by their parents. Three of thefe young birds were facrificed to my curiofity: The ftomach of the firft contained eight ftones, of the fecond eleven, of the third fifteen; together they weighed thirty-two grains, and confifted chiefly of quartz.

xxx. As thefe experiments did not anfwer my purpofe, it was neceffary to take up the enquiry at an earlier period, and make ufe of
ftill

still younger nestlings; nay, for greater certainty such as were just quitting the egg, and therefore could not have received food from their parents. The stomach, it is obvious, could not contain stones of any kind. I was therefore at the pains of keeping several nestlings in a warm place, while they remained unfledged, and feeding them till they were able to peck. They were then confined in a cage, and supplied at first with vetches soaked in warm water, and afterwards in a dry and hard state. In a month after they had begun to peck, hard bodies, such as tin tubes, glass globules, and fragments of broken glass were introduced with the food; care was taken that each wood-pigeon should swallow only one of these substances. In two days afterwards they were killed, when not one of the stomachs contained a single pebble, and yet the tubes were bruised and flattened, and the spherules and bits of glass blunted and broken: this happened alike to each body, nor did the smallest laceration appear on the coats of the stomach.

XXXI. I did not confine my observations to a single species. With the same view I set under a turky-hen several eggs, some her own, and some from a common hen. When the chickens were hatched I took charge of them.

them myself, and employed the same precautions as with the wood-pigeons (xxx). They were confined for fifty-five days in separate cages, and their food consisted of various sorts of grain. At last I forced them to swallow the hard indigestible substances so often mentioned. Upon examination, the stomachs appeared to be free from stones, yet the fragments and spherules of glass, and the tin tubes, were not on this account either the less or the more bruised or broken. Hence then we have at length a decision of the famous question concerning the use of these pebbles, so long agitated by authors. It appears, that they are not at all necessary to the trituration of the firmest food, or the hardest foreign substance, which is contrary to the opinion of so many anatomists and physiologists, as well ancient as modern; I will not, however, assert, that when put in motion by the gastric muscles, they are incapable of producing any effect on the contents of the stomach.

xxxii. But for what purpose are they designed? If they are not necessary to the trituration of the food, are we to suppose that they contribute in any other way to digestion? Do they create a keener appetite, or maintain a better state of health, as some conceive? Are they

they found in the ſtomach becauſe they are caſually mixed with the food, and as it were concealed by it; or, becauſe they are ſwallowed by choice, and even ſought after?

The firſt queſtions are already anſwered, or rather precluded, ſince we have found, that birds unprovided with pebbles take their food, are nouriſhed and grow juſt as well, and are as briſk and lively as others abounding with them; an obſervation I have made with ſatisfaction upon young pigeons, turkies, and chickens reared in the manner deſcribed above (xxx, xxxi).

xxxiii. The laſt queſtion will be readily ſolved, if grown up chickens take their food in the ſame way as young ones; for theſe ſwallow every thing that comes in their way. I have often ſcattered amongſt them various ſubſtances unfit for their nouriſhment, ſuch as pebbles, bits of brick, chalk, or other rubbiſh, which they pecked with eagerneſs, whether their ſtomach was full or empty. One day I threw among ſome chickens a large quantity of the little fiſh, termed *Lice* by Conchologiſts, which they devoured till their crop was full, juſt as if it had been the moſt agreeable food. If they retain the ſame diſpoſition when full grown, we may reaſonably conclude, that the collecting of pebbles is

leſs

less the effect of choice than stupidity; as the ostrich, according to Vallisneri and Buffon, devours without distinction whatever comes in its way, sticks, and stones, and cords, and glass, and metals, &c. such is its dulness, and so obtuse its organ of taste (*a*). But when fowls are grown to their full size, and when their natural instinct, which lay dormant while they were young, comes to be unfolded, change their manners in this as well as many other respects. A capon confined in a cage by Redi, died of hunger sooner than it would swallow pebbles offered to it in place of food (*b*). With me also three hens and a turkey, kept confined, died in the course of a few days, when instead of giving them food, I scattered before them a quantity of small stones. After their death, I found that the number of the stones was the same, though they would appear to be of the most proper kind, having been taken from the stomachs of other individuals belonging to the same species. When pebbles are mixed with the food, I have observed, that poultry especially when hungry, pick them up and swallow them. I should then incline to be-

---

(*a*) Buffon Hist. des Oiseaux. T. 2. Ed. in 12. Vallisn. Op. in fol. T. 1.

(*b*) Degli Anim. viventi negli Anim. viventi.

lieve,

lieve, that the stomachs of these birds generally contain a quantity of small stones; not because they are sought for and selected by design, as many suppose, but because they frequently happen to be mixed with their food.

xxxiv. Having shewn that the pebbles are not the cause of trituration of the food and other substances, we must conclude, that it is the sole and immediate effect of the gastric muscles, which, as it is well known, are very strong, and composed of firm and compact layers, and must, therefore, when set in motion, act with great force. To be more fully satisfied of this, let the stomach of a dog, sheep, or a man be compared with the gizzard of a duck, turkey, or goose; we shall then perceive the enormous difference between the thick muscular coat of the one, and the thin one of the other.

xxxv. The internal coat, or that which immediately lines the cavity of the stomach, deserves particular attention. In many animals, and in man himself, it is soft and villous; but in gallinaceous birds it is hard and cartilaginous. When separated from the next, which anatomists call the nervous coat, it soon becomes dry and very hard. In turkies and geese, in which it is thicker and stronger than in common fowls, I have

I have often diſſected it away, and ſpreading it upon a table, have drawn along it lancets, needles, bits of glaſs, and ſuch ſharp ſubſtances as are triturated in the ſtomach without any perceptible injury to it. If indeed I preſſed with conſiderable force, thoſe parts, over which the keen bodies paſſed, were diſunited, whether it was ſeparated or adhered to the other coats.

xxxvi. But theſe ſubſtances may act in a quite different manner when under the direction of the hand, than when ſet in motion by the gaſtric muſcles, and when the internal coat is not extended but forms a cavity, as it does when the ſtomach is entire. I therefore wiſhed to know what happens to ſubſtances incloſed in the ſtomach ſeparated from the animal, and preſſed externally with the palms of both hands, and agitated in various directions. The ſtomach of a turkey hen was firſt cleared of its contents by forcing them out through the pylorus, and then a large quantity of ſharp pieces of glaſs were introduced, which were kept in motion for a quarter of an hour by preſſure and percuſſion on the outſide of the ſtomach. I was in hopes, that I ſhould thus, in ſome meaſure, imitate the natural motion. Nor was the expedient altogether ineffectual; for the

internal

internal coat was only perforated with two little holes, fuch as the point of a needle would have made, and yet part of the glafs was reduced to powder, and part had loft its fharp edges. Different effects then are produced, when this coat is fubmitted to experiment after it has been removed from its natural fituation, and when it adheres to the others. Neverthelefs I am willing to allow, that how it fhould be capable of blunting and breaking the keeneft bodies without fuftaining any injury itfelf, ftill continues a matter of great furprize.

xxxvii. But if the infide of the gizzard be certainly agitated fo violently during the trituration of the food, will not the motion be perceptible on the outfide? Reaumur, induced probably by this reflection, laid open the abdomen of fome of the fowls in queftion, and watched the ftomach, but could not perceive what he perhaps imagined took place. They always feemed perfectly at reft, except the gizzard of a capon, which contracted and dilated alternately; he moreover faw certain flefhy cords moving in an undulating direction, but very flowly and gradually (*a*).

(*a*) Mem. cit.

xxxviii. I have perceived fimilar motions in two turkey cocks. Upon preffing the ftomach forcibly with my hand, I felt a flight pulfation that produced a fenfation of creeping, but was foon aware, that this was owing only to the beating of numerous little arteries, which run upon the furface of the vifcus. When a perforation is made in the heart of a living animal, and a finger introduced through it, it is well known that ftrong preffure is felt at the time of its contraction. I made this experiment upon the gizzard of a duck, but was not fenfible of the flighteft compreffion.

Conceiving that the ftomach muft exert its principal action when it is irritated by fubftances filling its cavity, I introduced fome nuts into the gizzard of a turkey-hen, that had been kept fafting for a day. During the whole time I watched it attentively, through an opening made in the abdomen; when it had received only a few nuts it fhewed no fign of motion, but when it was nearly full it fwelled violently, and then collapfed again of a fudden. Thefe alternations were fometimes general, and at others confined to a narrow fpace; they did not continue ten minutes, probably becaufe the aperture of the abdomen was bringing on the death of the

the animal. The nuts were unbroken, but evident contufions appeared upon their furface. This diftinct view of the motions of the ftomach I afcribe to unufual good fortune, fince, with the exception of only one other turkey, the ftomachs of many birds of the fame clafs remained at perfect reft, after they had been filled in the fame manner. If however we confider the very morbid ftate of the animal when the abdomen is laid open, we fhall not be much furprized at this phænomenon.

xxxix. The various facts related in the preceding paragraphs irrefragably prove, that the food of ducks, fowls, geefe, partridges, &c. muft undergo the mechanical action of the gaftric mufcles, before it can be broken down, and reduced to an impalpable pulp. But are we to fuppofe, that digeftion depends on this action, and that fimple trituration converts the aliment into that pultaceous mafs denominated *Chyme?* Or rather, that this mafs is generated by means of juices either prepared or collected in the ftomach; and that trituration is a co-operating, but not the immediate caufe of digeftion? I imagined that the tubes and fpherules, which had already afforded me fo much information, would not now be without their ufe. If the

gaftric

gaftric juices convert into chyme the food which trituration has prepared for digeftion, let fome food fo prepared be inclofed in the tubes and fpherules, and let us fee whether it will be diffolved according to this hypothefis; for then it muft be thoroughly foaked in thofe juices. I accordingly filled a tube and fpherule with crumb of wheaten bread mafticated, and introduced both into the gizzard of a hen. In three and twenty hours they were taken out, when the bread was much diminifhed in quantity, efpecially at the ends of the tube, where it was alfo fofter than at firft, and had acquired a bitter tafte. The fame tube and fpherule were forced into the gizzard of another hen, where they remained fourteen hours; after which there was no appearance of bread in either.

XL. I repeated this experiment upon a third hen, with bread of maize inftead of wheat; the tube and fphere were emptied in a day and half. As there was here no trituration nor any other power, except the action of the gaftric fluid, it feemed reafonable to conclude, that this fluid had diffolved and converted the bread into chyme, and fo enabled it to pafs through the holes in the receivers. A doubt however fuggefted itfelf, and kept me in fufpenfe; without fuppofing the

the tranfmutation of the bread into chyme, the gaftric fluid by merely diluting it, like water, might render it capable of paffing out of the tubes and fpherules.

XLI. A fubftance not foluble by fimple maceration, and at the fame time fofter than grain, upon which the gaftric juices have no action (III, IV, V, VI, VII.), was wanting to clear up the doubt. Flefh feemed to correfpond to this defcription. Flefh is digefted by many birds with gizzards, which for the moft part are both frugivorous and granivorous; I therefore filled four tubes with veal (*a*) bruifed very fmall in order to fupply the want of trituration, and forced them into the ftomach of a hen. They were taken out in twenty-four hours, and the flefh was in the following ftate: In the tube that came firft to my hands, it did not amount to above onetwentieth of its original bulk, in two others it had fuffered nearly the fame diminution; the only difference appeared in the fourth, which was not open at both extremities like the other three, but clofed at one end with a circular plate of iron. The flefh contiguous to the plate preferved its red colour and na-

(*a*) Wherever I mention flefh without an epithet, I mean raw flefh.

tural confiftence, and did not feem at all wafted; but at the open end it was reduced to two thirds of the length of the tube, of which it had at firft occupied the whole; the part that continued firm and red retained the true flavour of flefh; at the oppofite end it had entirely loft that flavour, and the furface, to the depth of a full line, was befides reduced to a pulp, and had acquired a cineritious colour. The inconfiderable refiduums in the other tubes were altered in the fame manner.

The immediate confequences of thefe experiments are felf-evident. The remarkable diminution of the flefh arofe from its having been in great meafure diffolved and digefted; for all phyfiologifts agree in confidering the change of colour and tafte, and the tranfmutation of the food to a pultaceous mafs in the ftomach, as the characteriftic marks of digeftion. The three tubes, of which the fides were perforated and the ends open, admitted the gaftric liquor at every part. Hence a confiderable wafte of the flefh in them. The cafe was different in the tube clofed at one extremity, and nothing can be more obvious than the reafon; for as the liquor could only enter at one end, it could only there diffolve the flefh.

XLII. This

XLII. This experiment decisively proves, that the gastric liquor was the cause of digestion in the present instance; and it was easy to foresee, that others upon the same class of birds would be attended with the same result. Some tubes filled with flesh were next introduced into the gizzard of a very large turkey cock, but the lattice work at the open ends, though it consisted of iron, could ill withstand the action of such powerful muscles. Upon examination seven hours afterwards, it was found separated from the tubes, and coiled up in one mass near the pylorus, in the midst of the pebbles and scoriæ of the food, some of which were jammed so tight in the tubes, that there was difficulty in forcing them out with the point of a penknife. I could not perceive the smallest fragment of flesh amongst them, and remained in doubt whether it had been digested, or expelled by these extraneous bodies. I resolved to submit this species of bird to further experiments, but was obliged to abandon the tubes, and have recourse to the hollow spherules, of which I have spoken above (VII). They were made thick and strong, with many small pores over the whole surface, in order to obviate two inconveniencies, the one lest the receivers should be unable to resist the violent

impulses

impulses of the stomach, the other to prevent the matters compressed and agitated by the action of the muscles from entering so readily into them. Two of these spheres were given to a turkey cock eleven months old, and in twenty-four hours were taken out of the gizzard. They contained at first about twenty-eight grains each of beef and veal bruised very small. Upon opening them after the same interval as before, and weighing the flesh, the beef was found to have lost nine, and the veal thirteen grains. I must not however omit to remark, that they were both fully impregnated with gastric liquor, and consequently would have weighed still less if they had been free from it. The beef and veal, when touched with the point of a knife, seemed tenderer than in their natural state, and resembled a soft paste rather than flesh. They had the bitter taste of the gastric juice with which they were impregnated, and the colour approached more to white than red. They were replaced in the sphere, and kept twelve hours in the gizzard of another turkey-cock. Upon a fresh examination, the beef weighed only eight, and the veal only five grains. The gastric fluid had therefore produced a new solution, and this process was entirely completed after the spheres, into which

which the flesh was put for the third time, had continued five hours in the stomach of a third turkey cock.

XLIII. Flesh is digested by the gastric liquor of geese as well as of turkies. Eleven grains of beef, inclosed in a spherule, were entirely dissolved in two days in the gizzard of one of these large birds.

I will not describe three other results obtained, one from an hen, and the two others from two capons; since, with respect to the digestion of the flesh, they are exactly like those just mentioned.

All these experiments were made with flesh bruised very small; this condition is not indeed indispensably requisite, but it very much promotes digestion. The bruised flesh was always dissolved in two days, but when entire that process was not completed in four, and sometimes not even in five days. The reason of this difference is obvious. The more flesh is bruised, the larger surface does it acquire; and in proportion to the increase of surface, more points are exposed to the action of the gastric liquor, which will consequently sooner complete the solution.

XLIV. Before I proceed further and conclude the present dissertation, I must notice an experiment of Reaumur, which does not perfectly

perfectly agree with thofe juft related. The greateft part of his memoir is employed in fhewing the great force of the gizzard of gallinaceous fowls in triturating the food; in the remainder he endeavours to prove, that this vifcus contains no menftruum of fufficient efficacy to produce folution. In fupport of this propofition, befides the argument derived from barley continuing unaltered within the tubes, he adduces the following fact, which requires to be particularly related. It is well known, how greedily ducks devour, and how foon they digeft, flefh. In order therefore to obtain the information he wanted, Reaumur had recourfe to this bird. Having provided fix tubes, four of lead, and two of tin, he inclofed in the former bits of veal of the fize of a grain of barley, and in the latter fome confiderably larger. Thefe fix tubes he gave to a duck at different times; viz. a leaden one at ten o'clock in the morning, and another at eight in the evening; next day a third was given at fix in the morning, together with the two tin tubes; laftly, at nine the fame morning the animal was made to fwallow the laft leaden tube, and at ten was killed. Of the four leaden tubes, one was voided the preceding day at nine in the evening; it was that which had been

taken

taken at ten in the morning; the other five remained in the gizzard, and the flesh was not only entire, but as firm as at first. Some of the pieces retained their red colour, three of them however had loft it. Of some of the tubes the whole capacity was no longer filled by the flesh; not that it had suffered any diminution, but because it was compressed by the stones and food, which had been admitted at the open ends of the tubes. From this experiment Reaumur infers, that no menstruum had acted on the flesh, since it was not either comminuted or diffolved. And though he does not affirm, that in the gallinaceous clafs digestion is the effect of trituration alone, he yet concludes, that the gizzard contains no folvent capable of decompofing and digesting the aliment.

XLV. What has been above related, shews how far Reaumur's conclufion ought to be extended; when we fpeak of aliment of a hard and compact texture, fuch as feeds, it muft be allowed, that the gaftric liquor has no action upon them (II, III, IV, V, VI, VII.); but when we are confidering food naturally tender, as flesh, or fuch as is made fo by art, as grain in the form of mafticated bread, it muft then be allowed, that a perfect folution is effected by the gaftric juices alone

(XXXIX,

(XXXIX, XL, XLI, XLII, XLIII.). In Reaumur's experiment the flefh remained fo fhort a time in the gizzard, that we cannot be furprized if it was not fenfibly diffolved. If we attend to the times at which he gave his tubes to the duck, and at which he killed it, we fhall immediately perceive, that the tube which continued longeft in the gizzard, remained in it only twenty-four hours; a fpace infufficient, according to my experiments on fowls, turkies, and geefe (XLI, XLII, XLIII.), for the gaftric liquor of thefe birds to diffolve any fenfible portion of flefh inclofed in tubes. I fhould however have condemned myfelf for a crime of omiffion, if, to the proof deducible from analogy, I had neglected to add direct experiments on ducks. Upon two ducks therefore I repeated the experiment of the French Naturalift, with the following variation; four tubes, each containing a bit of veal equal in fize to a barley-corn, were given to a duck; in two of the tubes the flefh was whole, but in the two others it had been previoufly cut into fmall bits: in fourteen hours the gizzard was examined; the four tubes were found in it; the two entire pieces of flefh were of their original fize, but inclining to a white colour; the fmall bits were alfo about the fame fize as at firft, but

were

were converted into a gelatinous paste. The experiment was repeated upon another duck, which was not killed till the end of the second day; and now the tubes that had contained the minute bits of flesh were entirely empty; and in the others, only some slight traces of a gelatinous concocted matter remained. If we combine these facts with others before related, it will appear, that in the gallinaceous class, trituration and the gastric fluid mutually assist each other in performing the important function of digestion; the former by breaking down the aliment, acts as the pre-disposing cause; the latter, when it is thus prepared, penetrates into it, destroys the texture, dissolves the particles, and disposes them to change their nature, and to become animalized.

XLVI. But what is the origin of this gastric fluid, so useful in digestion? How is it mixed with the food? And what successive changes does the latter undergo from the action of trituration, joined to that of the gastric liquor? These important questions required a strict examination of the œsophagus and gizzard, as also of the food during its passage through these parts, and continuance in them. As experiments are more conclusive, the greater the scale is, on which they are conducted,

ducted, I conceived that the larger species, as geese, turkies, ducks, and fowls, would be the beft fubjects for thefe enquiries. To begin then with the œfophagus of a goofe, this canal at the end towards the mouth, has the appearance of an inflated inteftine; it is above a foot long, and at its origin about an inch in diameter, but widens as it defcends, for the fpace of fix inches and more, when it contracts like a funnel, then enlarges again, and this enlargement continues to the gizzard. The œfophagus is membranous, its fides are ftrong and thick; they are thickeft at three inches diftance from the ftomach, on account of a flefhy fafcia, of which I fhall fpeak below. If we look very attentively, we can perceive the whole œfophagus covered with points or elongated fpots, which are moft numerous juft above the funnel. The fafcia appears to confift of a multitude of cylindrical bodies, fomewhat larger than hufked millet-feed. Thefe bodies are feen through a fine membrane, which furrounds the fafcia externally.

XLVII. If the œfophagus be inverted, and the fpots examined by the help of a glafs, we plainly perceive that they are follicular glands. This likewife is confirmed by the appearance of moifture on the œfophagus, when they are preffed.

preſſed. But the follicular glands that appear through the fleſhy faſcia like cylindrical bodies, bigger than millet, as we before obſerved (XLVI), are far more eaſily diſtinguiſhable, becauſe far larger. This faſcia, which encircles the œſophagus like a ring, is above an inch in breadth, and about a line in thickneſs. Great part of it is inveſted by a covering of a deep yellow colour, very thin and conſequently very liable to be torn. When this is removed, the faſcia externally appears white and rough, on account of the numberleſs prominent papillæ, each of which has a palpable pore in the center. When the faſcia is ſtretched, and ſtill more when it is preſſed between the fingers, a drop of whitiſh turbid liquor guſhes out at each pore into the œſophagus; and it may be increaſed, by continuing the dilatation or preſſure. The liquor is denſe, ſomewhat viſcid, of a ſweetiſh, and at the ſame time ſaltiſh taſte. To comprehend immediately that the pores are the excretory ducts of the follicular glands lying below, requires very ſlight anatomical knowledge: the glands appear very diſtinctly, when the membrane in which the pores are inſerted, is removed. The follicles are of a pale red colour, and full of a turbid liquor, which

which oozes out from the excretory ducts, when the œsophagus is kept under water.

XLVIII. Below the fleshy fascia, the œsophagus becomes membranous again for nearly the breadth of three quarters of an inch, when it is inserted into the gizzard. This organ is of the size of the fist, remarkably hard, and of an irregular elliptical figure; when opened lengthwise at the thinnest part, it is divided into two large muscles, each above an inch in thickness, and composed of very compact fibres. It appears plainly, that the whole action of these great muscles consists in approximating with violence, and like the sides of a vice, crushing and breaking to pieces all interposed substances. As the nervous coat adheres to these strong muscles, and as, however robust, it might be injured by such impetuous shocks, nature has sagaciously invested it with a cartilaginous coat, of a structure more capable of resistance, which internally lines the cavity of the gizzard.

XLIX. In turkeys the œsophagus and stomach very nearly resemble the same parts in geese. The former, however, is more membranous, and abounds more in follicular glands of a larger size, and consequently more conspicuous. The excretory ducts may be

easily

easily seen, and the liquor of the follicles may be readily forced out by pressure. This liquor is transparent, and somewhat viscid; its taste is rather sweet. But the œsophagus of the turkey has one peculiarity not found in the goose; it is provided with a bursa or bladder, well known under the name of the *crop* or *craw*. In this species it is very large. The crop at the sides at least, if not at every part, is furnished with follicular glands, exactly like the others. At the lower part of the œsophagus we also find the fleshy fascia, an inch in breadth, and provided with follicles much larger than those of the crop or œsophagus, and in great abundance. The liquor seems to have the same properties as in the goose. It is viscid, has a sweetish and saltish taste, a turbid white colour, and considerable density.

The gizzard, whether its form or the nature of its coats be considered, is exactly like that of the goose, only weaker and smaller in proportion to the inferior size of the bird.

L. I have observed all that has been related with respect to the gizzard and follicular glands of the goose and turkey in due proportion in the duck, common fowl, and even in smaller birds of the same class, as the pigeon, partridge, wood-pigeon, turtle-dove, and

and quail; with this peculiarity only, that in the duck the œfophagus, inftead of forming a crop, has the fame ftructure as in the goofe (XLVI). I fhall therefore omit a defcription of thefe parts, and proceed to confider the ftomach in a phyfiological light.

LI. In fpeaking of this organ, I have never mentioned either follicles or glands; for in the fowls hitherto mentioned, I could never difcover any. The internal coat, from its cartilaginous nature, appears to be unfit for the infertion of glandular bodies; at leaft I was not able to find the fmalleft veftige of them; nor did I fucceed any better in the nervous or mufcular coats, notwithftanding I examined them very narrowly. Reaumur having obferved a vaft number of fhort white filaments between the cartilaginous and nervous coats, entertained fome fufpicion of their being tubes or veffels, placed there in order to difcharge their contents into the ftomach (*a*). I have found thefe filaments in all the gallinaceous fowls I have examined; but cannot agree with him that they remain attached to the nervous, when the cartilaginous coat is feparated from it: for after fuch feparation, I have ever feen them adhere to

(*a*) Mem. cit.

the cartilaginous, never to the nervous coat; but any perfon may readily make the trial. Thefe filaments are very numerous; they are pointed at the extremity, oppofite to that which is inferted into the cartilaginous coat, and refemble fhort white down, diftinctly vifible by the naked eye in the larger birds; fuch as the goofe and turkey, but requiring the aid of a glafs to be feen in the fmaller fpecies. I have divided many of various fizes with the points of very fine needles, in order to difcover whether they were hollow or glandular, but could never find any appearance of this kind: I have alfo fqueezed them in order to fee if any liquor would ooze out, but to no purpofe: and fo far from fufpecting thefe filaments of Reaumur to be vafcular or glandular, I fhould rather fuppofe them to be merely for the purpofe of joining, or at leaft more clofely connecting the cartilaginous with the nervous coat.

We fhall fee elfewhere, that ftomachs of the membranous kind, when they are taken out of the animal and rubbed dry, foon become moift again: this moifture comes from invifible veffels and glands, difcharging their liquor into the cavity of the ftomach. I have made the fame experiment on mufcular ftomachs, but they always continued dry; the
fame

same thing alfo took place, when I preffed them underneath, though this is a very effectual means of accelerating and increafing the covering of moifture. Hence I have good reafon to fuppofe, that the juices found in mufcular ftomachs do not properly belong to them, but come chiefly from the œfophagus, and in part from the duodenum, as we fhall fee below.

LII. Nature however, has not failed to provide the quantity neceffary for digeftion. We have feen the vaft number of follicular glands with which the œfophagus is provided (XLVI, XLVII, XLVIII, XLIX.); they muft needs pour in their liquor in great abundance. And experience confirms what reafon fuggefts. I introduced a fmall piece of dry fpunge, previoufly cleanfed from every impurity, into the craw of a pigeon, in which it was left twelve hours; at the expiration of which time I opened the craw, and took it out. The fpunge was full of liquor, and on being fqueezed into a glafs, afforded above an ounce. I employed larger pieces of fpunge in fowls and turkeys, and obtained more of this œfophageal liquor; the quantity in a turkey amounted to feven ounces in ten hours. A fimilar liquor is procured in equal abundance, from fuch œfophagufes as are dilated into a

large

large canal, inftead of a craw, as in ducks and geefe (XLVI, L.). This fluid is undoubtedly defigned to foften the food which remains a certain time in the craw, or in the large canal; which not only difpofes it to be more readily broken down, but very probably alfo communicates to it fome quality that renders it more eafily digeftible. But it is likewife certain, as I have found from experiment, that a confiderable part of this fluid defcends into the ftomach; not to mention that denfer and more vifcid liquor which diftils from the flefhy fafcia, lying at the bottom of the œfophagus, (XLVI, XLVII.).

LIII. Thefe various œfophageal juices acquire in the ftomach a bitter flavour, refembling that of the food in this vifcus: and as this tafte exactly refembles that of the bile, which in thefe animals is difcharged through the cyftic duct into the duodenum, I am thoroughly perfuaded that it arifes from this fource, in confequence of the bile regurgitating into the cavity of the ftomach, and being mixed with the food and œfophageal liquors collected there. I am confirmed in this perfuafion by other facts, for the relation of which I fhall find a more convenient place; not to mention the well known circumftance

of

of the bile being found in the ſtomach of various animals (a).

LIV. This collection of divers liquors in the gizzard of our fowls, ſerves as a menſtruum for the food, and diſpoſes it to be tranſmuted into chyle. But the firſt ſtep towards this event is taken in the craw. It is there that the aliment is penetrated by the œſophageal liquor, and begins to change its ſmell and taſte: that of the hardeſt texture is prepared to be broken down when it diſcends into the ſtomach, which in theſe birds may be ſaid to ſupply the place of teeth.

But the way in which the food deſcends from the mouth into the ſtomach, is deſerving of attention. When our fowls are abundantly ſupplied with meat, they ſoon fill their craw: but it does not immediately paſs hence into the gizzard, nor does not arrive there till after it has been macerated in the craw: it always enters in very ſmall quantity, in proportion to the progreſs of trituration in the ſtomach. Here then, what happens in a mill, may be obſerved to take place. A receiver is immoveably fixed above the two large ſtones which ſerve for grinding the corn;

(a) Haller Elém. Phyſiol. T. 6. Valliſn. Op. in Fol. T. I.

this receiver lets the corn which it contains, fall continually in small quantity into the central hole in the upper stone, through which it passes, and diffuses itself in the void space between the two stones, where it is broken down, triturated, and pulverized by means of the strong friction of the upper stone that moves round with great velocity upon that below. Meanwhile the flour passes from between the stones, as substances triturated by the gizzard, and dissolved by the gastric juices, are expelled through the pylorus into the small intestines.

LV. All this may be observed, by inspecting the alimentary canal during the time of digestion. If the bird has fed upon grains, they are found in the cavity of the gizzard, partly entire, but softened by a fluid. That part of the œsophagus that lies between the end of the crop and the beginning of the stomach, either contains no grains at all, or only a few quite entire. Trituration takes place in the gizzard only. Those which have first entered this cavity, are found to have lost the farinaceous substance, and are reduced to mere bran; the succeding ones are more or less broken, and the last are entire. Amid this mixture of bran, and broken and entire grains, we always find a semi-fluid pultaceous

taceous mafs of a whitifh yellow colour. This is the farinaceous parts of the grains decompofed by the gaftric liquor, and converted into chyme. Meanwhile frefh grains continue to fall into the gizzard, in order to undergo the fame tranfmutation: this admirable procefs continues as long as any grains continue to fall into the ftomach.

Thefe appearances and changes take place alfo in animal fubftances, whenever birds with mufcular ftomachs feed upon them.

LVI. At whatever time the ftomachs of thefe birds happen to be opened, they always contain a certain quantity of gaftric liquor. But it is lefs abundant when they are full of food, (being in this cafe abforbed by the food) than when they contain little or none. If we wifh therefore to be provided with a large quantity of this liquor for experiments, it fhould be taken from the empty ftomach. Befides, in this cafe it is purer than when mixed with the food. When examined in a ftate of purity, its tranfparency, if we except a flight yellow tinge, is little inferior to that of water. It has likewife the fluidity, but not the infipidity of water, being always a little bitter, as well as falt. I have found that the gizzards of turkeys and geefe abound moft in gaftric juices, probably on account of their

their superior size. I was induced by the quantity they afforded to attempt an experiment, which if it succeeded, would still further prove that trituration is only an assisting or predisposing, and not the efficient cause of digestion. It consisted in trying, whether these juices retain their solvent power out of the stomach. For this purpose, I took two tubes sealed hermetically at one end, and at the other with wax: into one I put several bits of mutton, and into the other several bruised grains of wheat, and then filled them with the gastric liquor. In order that they might have that condition which in these animals precedes digestion, they had been macerated in the craw of a turkey cock. And as the warmth of the stomach is probably another condition necessary to the solution of food, I contrived to supply it by communicating to the tubes a degree of heat nearly equal, by fixing them under my axillæ. In this situation I kept them at different intervals for three days, at the expiration of which time I opened them. The tube with the grains of wheat was first examined; most of them now consisted of the bare husk, the flour having been extracted, and forming a thick grey sediment at the bottom of the tube. The flesh in the other tube was in great measure

sure dissolved, (it did not exhale the least putrid smell) and was incorporated with the gastric juice, which had become more turbid and dense. What little remained had lost its natural redness, and was now exceedingly tender. Upon putting it into another tube, and adding fresh gastric liquor, and replacing it under the axilla, the remainder was dissolved in the course of a day.

I repeated these experiments with other grains of wheat bruised and macerated in the same manner, and likewise upon some flesh of the same kind, but instead of gastric juice I employed common water. After the two tubes had remained three days under my axillæ, I found that the grains, where they were broken, were slightly excavated, which was occasioned by an incipient solution of the pulpy substance. The flesh had also undergone a slight superficial solution, but internally it appeared fibrous, red, firm, and in short, had all the characters of flesh. It was also putrid; the wheat too had acquired some acidity, two circumstances, neither of which took place in the grains and flesh immersed in the gastric liquor. These facts are then irrefragable proofs that the gastric juice, even out of its natural situation, retains the power

power of diffolving animal and vegetable fubftances in a degree far fuperior to water.

LVII. The gaftric juice which I employed was taken from a turkey. That of a goofe produced fimilar effects. I have further found, that in order to operate the folution of animal and vegetable fubftances, this juice fhould be frefh. It lofes its efficacy, when it has been kept fome time in veffels, efpecially if they fhould happen to be open. It alfo becomes inefficacious after it has been ufed for one experiment. Laftly, a confiderable degree of heat, equal to the temperature of man or birds, muft be applied; otherwife, the gaftric juices are not more effectual in diffolving flefh and vegetables than common water. This artificial mode of digeftion is well calculated to illuftrate the fubject I have undertaken to treat; but I fhall have opportunities of fpeaking of it at greater length in the fubfequent differtations.

DISSERTATION II.

CONCERNING THE DIGESTION OF ANI-
MALS WITH AN INTERMEDIATE STO-
MACH, CROWS, HERONS.

LVIII. BY the term intermediate stomach, I mean such a stomach as, on the one hand, is not properly muscular, that is, provided with thick and strong sides, as in the gallinaceous family (1.); and on the other, is not merely membranous, that is, very thin, as in birds of prey and man, but has an intermediate degree of thickness and strength. The stomachs of both the raven (*a*) and grey crow (*b*) may be considered in this light, though in reality they approach nearer to the muscular than the membranous class.

(*a*) These two species are called by Linnæus, Corvus cine-
rascens, capite, jugulo, alis caudaque nigris.
Corvus ater, dorso atro-cœrulescente, cauda subrotunda.
(*b*) The hooded crow of Pennant. Corvus Cornix L.

The

The intermediate power of these stomachs contributes also to characterize them; it is very far from being equal to the force of muscular, but greatly exceeds that of membranous stomachs. Such tubes of tin, as doves and pigeons would flatten and disfigure with the greatest ease, remain unaltered in the stomach of crows. Thus also grain is triturated by the former, but continues whole in the latter. Their gastric muscles however are not inert. They exert a certain degree of action, but it is far inferior to that of the gizzard in the gallinaceous class. Thus, though they cannot compress tin tubes, they are capable of producing this effect upon tubes of lead, provided they are very thin: and those that continue unaltered at first, are at length slightly incurvated or distorted at the edges, and generally filled with fragments of the food, evident marks of considerable action in the gastric muscles; there are no effects which shew such action in animals with membranous stomachs, as we shall find in its proper place. I have often seen these phænomena, having kept a great number of grey crows and ravens, which have been very serviceable in the course of my enquiries, as the reader will learn from a perusal of the present dissertation.

LIX. These

LIX. These birds may, as well as man, be denominated *omnivorous*. Herbs, grafs, feeds of leguminous plants, flesh of every kind, alive or dead, serve equally for their nourishment. As the powers for the concoction of various aliments, possessed by these species are either entirely the same as, or strongly resemble those of man, it is obvious, that the knowledge obtained from them will greatly illustrate the process of digestion in us. They besides seem formed on purpose to forward the views of the observer. When we wish to know what changes have been produced in substances inclosed in spheres and tubes, and given to gallinaceous birds, it is necessary to extract the tubes and spheres from their gizzards; that is, it is necessary to kill them. Hence for every experiment we must sacrifice an individual, at no small expence to our philosophical curiosity. On the contrary, we can perform such experiments upon crows as often as we please, without destroying a single individual. With respect to substances they are incapable of digesting, such as the above-mentioned metallic receivers I have discovered, that they possess the privilege of returning them through the mouth, as birds of prey vomit the feathers and hair of the animal they have devoured, a circumstance well

known

known both to naturalifts, and thofe who train falcons for the field. But whereas this vomiting generally takes place every twenty-four hours in birds of prey, in crows it happens at leaft every nine, and commonly every two or three hours.

LX. As I obtained the fame refults from both fpecies, I fhall employ in my narration the generic name only (*a*). My obfervations were begun in winter, the moft convenient feafon for procuring a large number, owing to the multitudes, efpecially of ravens, with which Auftrian Lombardy, and indeed almoft all Italy then abounds. All the crows which I could obtain, had when newly taken, a large collection of pebbles in the ftomach; the biggeft were of the fize of fmall peafe, the leaft of that of millet: they were of various forts; I even found rounded pieces of brick. But in lefs than ten days not a ftone remained in the body, a circumftance which I learned from the infpection of feveral ftomachs, when I had occafion to kill fome crows in order to obferve the anatomical ftructure of the alimentary canal. They were voided partly at the anus, as appeared from the excrements,

(*a*) Corvus is the generic name in Latin, and Cornachcia in Italian, and Crow may very well ferve for it in Englifh.

and

and partly by the mouth; in the latter inſtances they were glued by the gaſtric liquor to the outſide of the tubes which I had forced them to ſwallow, and which they afterwards threw up. When unprovided with pebbles, they continued to eat and were nouriſhed as well as before. Hence it is to be inferred, that they are not more neceſſary to digeſtion in birds with intermediate, than in thoſe with muſcular ſtomachs, as we have ſeen above (xxxi). And as I inclined to believe, that the laſt mentioned claſs do not pick up theſe ſtones from choice, but by mere accident (xxxiii); ſo I conſider the matter likewiſe with reſpect to crows, having obſerved, that though unprovided, they never peck them eagerly, even when hungry, but ſwallow them only when they happen to be mixed on purpoſe or by chance with their food, and as it were concealed by it.

LXI. I began my experiments by putting whole beans or grains of wheat in the tubes (*a*). The reader will eaſily perceive, that crows are not ſo ſtupid as to take the tubes ſpontaneouſly, but that it is neceſſary to force them down the throat, and to paſs the finger

---

(*a*) Theſe tubes were the ſame I uſed for gallinaceous fowls, and I continued to employ them in the ſequel.

along

along with them till they are got into the ſtomach. This I executed in the way I had before done in animals with muſcular ſtomachs (111). The tubes were all thrown up in the ſpace of three hours. The beans and wheat appeared as at firſt, excepting that they were ſomewhat ſoftened and ſwelled by the gaſtric juice, which had penetrated a little way into them. I replaced the grains in the tubes, and introduced them again into the ſtomach, where they remained two hours longer, without undergoing any further change. I repeated the ſame experiment a great number of times, and upon computing the ſpace during which the tubes had continued in the ſtomach, I found that it amounted to forty-eight hours; in this interval the feeds had ſuffered no other alteration, except in being a little more moiſtened. The gaſtric fluid is therefore incapable of effecting the ſolution of theſe vegetable matters.

LXII. But we have before ſaid, that they were entire; on which account, it could not act upon the farinaceous ſubſtance of the grain till it had traverſed the huſk; and this might have diminiſhed its efficacy. In order to determine how far this ſuſpicion was well-founded, it became neceſſary to repeat the experiment upon the ſame feeds bruiſed.

Accord-

Accordingly four tubes full of the coarfe flour were given to a crow: they remained eight hours in the ftomach, and proved the juftnefs of my fufpicion; for upon examining the contents, I found above a fourth part wanting. This could arife from no other caufe but folution in the gaftric liquor, with which the remainder was fully impregnated. Another obfervation concurred to prove the fame propofition: the largeft bits of wheat and bean were evidently much diminifhed; this muft have been owing to the gaftric liquor having corroded and diffolved good part of them, as the nitrous acid diluted with a large quantity of water, gradually confumes calcareous fubftances. I replaced what remained of the feeds in the tubes, and introduced them again into the ftomach, wherein they remained, at different intervals, twenty-one hours; when they were entirely diffolved, nothing being left but fome pieces of hufk and a few inconfiderable fragments of the feed.

LXIII. Wheat and beans floating loofe in the cavity of the ftomach, undergo the fame alteration as in the tubes. When I fed my crows with thefe feeds, I obferved, that before they fwallowed them they fet them under their feet, and reduced them to pieces by repeated ftrokes of their long and heavy beaks.

And now they digested them very well; nay, this procefs was very rapid in comparifon of that which took place within the tubes. But when the birds either from exceffive hunger or violence fwallowed the feeds entire, the greateft part of them paffed out entire at the anus, or were returned by the mouth. We cannot therefore be furprized, that the gaftric juice could not diffolve them within the tubes, fince it was incapable of effecting this procefs within the cavity of the ftomach, where its folvent power is far fuperior.

LXIV. To avoid prolixity, I fhall not fpeak of other feeds fubmitted to the fame experiments; fuch as chicken-peafe, French beans, peafe, and nut-kernels. I will rather mention vegetable matters of a fofter and more yielding texture, which did not require to be broken down in order to be diffolved; fuch as crumb of bread and apples. Thefe fubftances were not only diffolved within the tubes, but required a much fhorter time than beans and wheat. Several bits of a ripe apple, weighing together eighty-two grains, and inclofed in tubes, were diffolved in the fpace of twenty-four hours in the ftomach of a crow. Four bits of another apple, weighing an hundred and three grains, were diffolved in little more than fifteen hours. Of an an hundred and seven grains of crumb of wheaten bread, there only remained eleven in the space of thirteen hours.

LXV. From vegetable I proceeded to animal substances. The eagerness which crows shew for these afforded a certain presage, that they would be dissolved within the tubes. I filled eight tubes with beef, and gave them to four crows, two to each. The flesh was not bruised small, as in the case of gallinaceous fowls (XLII), but each tube contained a whole piece. An hour had scarce elapsed when one was thrown up. The flesh, upon examination, did not appear to be sensibly diminished, but it was thoroughly soaked in the gastric liquor. The juice was a little bitter, and of a yellowish green colour; the flesh had acquired the same taste and colour in several places. In an hour and three quarters two other tubes were vomited up; and now the flesh began to shew marks of solution. The red colour was changed to a dark cineritious hue, and the whole surface was become flabby, and the cohesion of the parts was destroyed. In another tube, discharged in two hours and an half, the solution had made a greater progress. A dark covering of jelly surrounded the flesh, which on being touched, adhered to the fingers; and when applied to the tongue,

tongue, hardly exhibited the flavour of flefh. The folution had proceeded ftill further in four hours, when two other tubes were thrown up; in which the flefh did not amount to half the original quantity. The remainder was furrounded by the fame gelatinous covering, under which it preferved its natural colour, fibrous ftructure and favour. There remained only two tubes, which were vomited up feven hours after they had been taken. Both were empty, the flefh therefore had been completely diffolved, except a few bits of jelly that adhered to the infide. I never could perceive the fmalleft token of putrefaction either during the progrefs, or at the completion of the folution. And this obfervation, that I may not be under the neceffity of repeating it continually, is to be extended to all the folutions performed not only by other crows, but by all the animals that I fhall have occafion to mention in this work; for I can affert with the utmoft confidence, that I have never been fenfible of the flighteft ftench either in flefh or any other fubftance which I introduced in tubes or any other way.

Nothing could be more fatisfactory than the information obtained from this experiment. It not only rigoroufly demonftrates, that the gaftric liquor of crows is the folvent of of flesh inclosed in tubes without borrowing the least aid from trituration, but it throws still stronger light upon the mode of operation of this menstruum in the gallinaceous class. It begins by softening the texture and altering the colour; next succeeds the de-composition of the parts; this transmutes the flesh into a kind of jelly of a taste different from that of flesh: the jelly is then more thoroughly penetrated by the juice and extracted out of the tubes, and in the stomach it is changed into chyle. It appears also, that this fluid does not penetrate deeply into the flesh, but acts on the surface only, dissolving and removing one layer at a time, if we may so speak, like other corroding menstrua, until it comes to the innermost part, which it also softens and melts.

LXVI. We have seen, that the flesh in the tubes shewed no sign of solution till an hour and three quarters had elapsed, and that this process was completed at the end of seven hours (LXV). But are we to conclude, that this is the measure of the time required by the gastric liquor for this operation? or that it would have been accomplished in a shorter time, if the liquor had had free access to the flesh? for it is certain, that the tubes are no small impediment to the gastric juice. What then

then would happen if the impediment was in part removed? and what when it is entirely taken away, by putting the flesh loose into the stomach? In order to solve the first of these interesting questions, I enlarged the perforations in the sides of the tubes as much as possible (VII), then filled them with beef, as before (LXV), and introduced them into the stomachs of several crows. I now had the pleasure to perceive the superior efficacy of the gastric liquor. In an hour and an half three of the tubes were thrown up, and above a fourth of the flesh appeared to be wasted. Two other tubes were discharged in less than two hours, and contained little more than half of their original quantity. And before the completion of the fourth hour, the remaining tubes were entirely empty.

LXVII. Before I proceeded to the other question, I thought of inverting the foregoing experiment (LXVI), and instead of allowing freer access to the gastric juice, of impeding it more and more, and at last hindering it almost entirely. I began with employing the usual tubes wrapped in cloth; this, although it was thin, was sufficient to prevent the solution of the flesh, which now did not begin to take place till three hours after the tubes were introduced into the stomach,

mach, and was not completed till ten had elapfed.

The linen in which the tubes had been wrapped was fingle: I now doubled it in order more effectually, to hinder the ingrefs of the liquor, and repeated the experiment in the fame manner. The flefh fhewed no token of folution for four hours, and was not entirely diffolved for a whole day.

Upon wrapping round another fold the folution did not begin for nine hours, and in the fpace of a day the flefh was fcarce half confumed. In other refpects the gaftric juice had acted upon the flefh juft as it does in open tubes, excepting only the flownefs of its operation. It was become externally gelatinous, and incoherent in its parts. It was tinged yellow in feveral places; the tafte and fmell at the furface were not different from thofe of the gaftric liquor.

I concluded thefe experiments, by trying what would be the effect of putting flefh into tubes with only three or four holes. After they had continued nine hours in the ftomach, the refult was as follows: fmall excavations of greater or lefs depth were made in the parts oppofite to the pores, and from thefe excavations fmall furrows wandered irregularly along the furface of the flefh. The flefhy fibres

fibres both in the cavities and furrows were become exceedingly tender, they had befides loft their red colour, and were turned yellow. The reft of the flefh was unaltered. From what has been faid before, the origin of the cavities and furrows evidently appears to have been derived from the gaftric juice, which by infinuating itfelf through the little perforations, had there diffolved and deftroyed the flefh; the reft remained entire, becaufe none of the juice could enter, if we except a very flender ftream, which had produced the furrows.

LXVIII. Let us now proceed to the fecond queftion and examine how much more readily flefh lying loofe in the ftomach is digefted than when it is enclofed in tubes. Taking fome of the fame kind of flefh that had been ufed before, viz. beef, I parted it into two equal portions, one of which was again divided into fmaller bits before it was put into the tubes, while the other was left entire. Each portion weighed eleven pennyweights. I next gave the tubes, which were eight in number, to a crow, and to another bird of the fame fpecies, equally healthy and robuft, I gave at the fame time the whole portion of flefh, to which I had previoufly faftened a thread. This thread, hanging out of the bird's mouth, and

and being wrapped round the neck, I could draw up and examine the flesh at pleasure. And that every circumstance might be alike, I had taken care that the two crows should have their stomachs empty. In thirty-six minutes one of the tubes was vomited, and at the same instant I drew up the flesh from the stomach of the other crow. The latter was throughly imbibed with gastric juice, especially the part that rested upon the bottom of the stomach. It had lost its redness, and was now of a dirty colour; it weighed forty-two grains less than at first; on the contrary, the flesh enclosed in the tube retained its original weight.

The tube and the flesh tied to the string, were replaced in their respective situations; and in order that both might remain the same length of time in the stomach, I took care to return the tubes as they were thrown up. The flesh was entirely dissolved in three hours, when I immediately killed the crow that had the tubes. Upon collecting and weighing all the flesh that remained in them, I found it to amount to about seven pennyweights. Hence in three hours and nine minutes it had lost four pennyweights.

On the other hand, the flesh tied to the string was reduced to half a pennyweight, which

which confifted of a packet of membranous or cellular fibres, the flefhy part having been entirely diffolved. This experiment clearly fhews, that flefh left loofe in the ftomach is more fpeedily digefted than when it is enclofed in tubes. And theory perfectly agrees with fact; for fince folution is the effect of the gaftric fluid, it is evident that the food, when loofe in the ftomach, is attacked by a larger quantity than when defended by the tubes.

LXIX. Young crows, as well as all other young birds, eat more than the adult; hence I fufpected their digeftion to be quicker. Having a neft of the grey fpecies brought me in June, I made, among others, the experiment related in the laft paragraph. The refult was very fatisfactory. A quarter of an ounce of beef, faftened as before, to a thread, had fcarce touched the ftomach, when the folution began, and in forty-three minutes was completed; but an equal quantity diftributed in feveral tubes, required four hours and a half to be diffolved. Upon opening the ftomachs of the two young birds, I immediately perceived the caufe of this rapid folution; they contained half a fpoonful of gaftric fluid; a quantity feldom met with in the ftomach of adult crows. As the neftlings require more food, Nature has furnifhed them

them with the means of an eafier and more fpeedy digeftion.

It is fcarce neceffary to remark, that the experiments related in the Lvth and following paragraphs, clearly evince this important truth, that the digeftion of food is proportional to the quantity of gaftric juice acting upon it. When this liquor comes in contact only with a few points, the decompofition is very flow and inconfiderable (LXVII); when freer accefs is allowed, the folution takes place more fpeedily, and is more confiderable (LXV, LXVI); it is very rapid, when every obftacle is removed, and the food is on all fides expofed to the action of the folvent liquor, (LXVIII, LXIX).

LXX. It is a queftion of ancient date, and ftill agitated by modern phyfiologifts, whether certain carnivorous animals are capable of digefting bone. Among the various points, which I propofed to difcufs in the prefent work, I conceived that this well deferved the reflection and attention of the philofopher; I fhall therefore both here, and in another part of my work, relate what I have obferved on the fubject. If we obferve a crow and a bird of prey devouring an animal, we may be difpofed to think that the latter has the power of diffolving bone, but not the former.

When,

When, for inftance, a hawk takes a pigeon, it firft ftrips the back, and devours the mufcular part of the breaft; then proceeds to the entrails; and, laftly, fwallows the ribs, vertebræ and head, not even fparing the feet and wings, if it fhould happen to be very hungry. When the fame bird is given to a crow, it fets about ftripping off the flefh; but when it has picked this clean, it leaves the fkeleton. This rejection of the bones, is however very far from being an indubitable proof, in the eftimation of the philofopher, that they are incapable of digefting them. At moft it inclines us to believe it probable; but fuch probability requires to be confirmed by facts: and being engaged in enquiries of this nature, I could conveniently bring the queftion to the teft of experiment. As I happened to be provided with fome phalanges of the human toes, I enclofed two in one of the ufual tubes, which remained thirteen hours in the ftomach. They weighed fifteen pennyweights at firft; nor was this weight at all diminifhed, or the bones in the leaft foftened. Being in doubt whether the two great thicknefs of thefe bones might not have prevented the gaftric juice from acting upon them, I had recourfe to fmaller ones. Happening one day to find one of my crows dead in the apartment where I kept them, and the reft affembled round the carcafe in crouds, and devouring it eagerly, I took one of the tibiæ, broke it in two, and enclofed it in a tube. The tube continued a whole day in the ftomach of another crow, but the bone was neither foftened, nor diminifhed in weight. The fame thing was alfo obfervable, after the bone had been left loofe in the ftomach for fourteen hours longer.

LXXI. The greedinefs with which the crows devoured their companion, induces me to digrefs, for the fake of noticing a miftake of the celebrated Dr. Cheyne. He pretends, that crows cannot digeft the flefh of their own fpecies; and that when they happen to fwallow it, they vomit it up again. " Ipfa Cornix (fays Haller, on the authority of Cheyne) cornicis canem ingeftam non poteft coquere & deglutitam vomitu rejicit (*a*)." But the truth is, that the flefh which my crows devoured agreed very well with them, nor did they throw any of it up again. Further, in order to determine certainly whether the above-mentioned writer had fallen into a miftake or not, I killed and plucked another crow, and threw it into the chamber where

(*a*) Phifiol. T. 6.

its companions were kept, when they immediately leaped upon it, and devoured it with the fame avidity as they had done the other, without afterwards vomiting the leaft particle of it. Upon killing and opening, three hours afterwards, one which appeared to have loaded its ftomach more than any of the reft, I found the flefh partly diffolved, and in the form of a femifluid pulp, and partly in the procefs of folution, the very ftate in which I had feen other flefh.

LXXII. But let us return to bones. It appears that, whether large or fmall, they are alike infoluble in the gaftric juices of crows (LXX). But is this true likewife with refpect to thofe of a foft ftructure, and which bear fome refemblance to cartilage? In order to afcertain this point, I made ufe of another tibia, taken from an unfledged crow, and which therefore had not acquired its natural rigidity, though it was fo hard as to break, when I tried to bend it: and now the gaftric liquor was not inactive. Of fifteen grains, which it weighed at firft, it had loft five, after continuing fix hours in the ftomach enclofed in a tube. It was become fo foft, that it was capable of being bent into the fhape of a bow. It continued to wafte and become fofter; and when it had remained twenty-feven

ven hours in the ſtomach, it was ſo much reduced as to reſemble a thin tube of paper. It was not at all gelatinous; and when it was preſſed between the fore-finger and thumb, it ſhewed ſome elaſticity, by recovering its former ſhape when the preſſure was removed. It was not ſcabrous either internally or externally, but had rather acquired a greater degree of ſmoothneſs during its ſolution. In five hours more it had loſt the ſhape of a tube, and was totally reduced to pieces.

LXXIII. I tried other tender bones belonging to larger animals; and more or leſs of them was diſſolved, but with difficulty, and after a very long interval. The ſolution was more ſpeedy in young crows, probably on account of the greater abundance of their gaſtric juices (LXIX).

With reſpect then to the queſtion concerning bones, we muſt conclude that they are indigeſtible by crows, except only ſuch as, on account of their ſoftneſs, are rather to be conſidered as cartilage than bone.

LXXIV. In the preceding, as well as the preſent diſſertation, I have always ſpoken of the ſtomach as the place deſtined for the concoction of the food. And in truth, whether we conſult antient or modern phyſiologiſts, or conſider my experiments, it will appear ſo

clearly

clearly proved, that it would be abfurd to entertain a doubt of it. But it may be proper to enquire whether this operation belongs exclufively to the ftomach in the birds in queftion, or is partly carried on in the œfophagus. This enquiry is fuggefted upon the manifeft decompofition, which has been obferved in part of the food that is found in the œfophagus of fome animals, as among others in the fea-crow and the pike (*a*). In order therefore to afcertain this point, I was led to make a few experiments, which I fhall relate after I have given a fhort defcription of the œfophagus and ftomach in crows, and of the fources of the refpective liquors in thefe two cavities.

LXXV. The œfophagus is membranous, and has no craw. When dilated it is cylindrical, if we except a flight contraction in the middle. To the naked eye it would feem deftitute of follicular glands, which however become confpicuous when it is viewed with a glafs. They are in fuch abundance, that there is not a fingle point of this canal without numbers of them. The excretory ducts are fcarce difcernable, though they emit the liquor of the follicles in great plenty. To fee

(*a*) Helvetius Mem. de l'Acad. 1719. Plot Nat. Hift. of Staffordfhire.

this, it is sufficient to pass the finger over them. The liquor is of a viscid nature, of a cineritious white colour, and somewhat sweetish to the taste.

The inferior part of the œsophagus has the same kind of fleshy fascia that has been noticed in birds with muscular stomachs. This fascia in crows is scarce an inch long; and in them too, as well as in the class of birds just mentioned, is a tissue of large follicular glands very evident to the naked eye, of a roundish figure, and full of a sweet fluid, less viscid than that in the small follicles in the membranous part of the œsophagus, but more dense, and of a lighter cineritious hue.

LXXVI. In the gallinaceous tribe we have spoken of three coats, the cartilaginous, nervous, and muscular (XLVIII, XLIX), which principally compose the stomach. These three coats are likewise found in birds with an intermediate stomach. When the cartilaginous is separated from the nervous coat, and the latter is viewed with the naked eye, it appears to contain a multitude of whitish little bodies inchased in it, which have the appearance of points; but when examined by the microscope, change their appearance to that of follicular glands, much smaller than those in the fleshy fascia (LXXV); these follicles

licles are full of a vifcid liquor, which they difcharge at the extremity towards the ftomach, when they are preffed by the finger, or any other body. The difcovery of thefe glands in the nervous coat having led me to imagine, that they might empty their contents into the ftomach, I examined the cartilaginous coat with great attention, in order to try if I could find any minute pores for the tranfmiffion of the liquor into the cavity of that vifcus; but I acknowledge ingenuoufly, that I could not difcover any. This however by no means proves their non-exiftence; for they may be fo fmall as to elude the fight, even when aided by the microfcope. And I cannot but believe, that thefe follicles, of which the excretory ducts are turned towards the ftomach, are deftined by nature for pouring their contents into that organ.

LXXVII. I now proceed to enquire whether, exclufively of the ftomach, digeftion is at all performed in the œfophagus of crows. In order to determine this, I firmly fixed to an iron wire two equal pieces of veal, one of them to the end of the wire, and the other two inches above. I then forced it down the throat of an hungry unfledged crow; the piece faftened to the end lay in the ftomach, while the other occupied the œfophagus.

To

To prevent them being thrown up, a ftring was faftened to the upper end of the wire, brought out at the mouth, and tied round the beak. Thus I was enabled to draw up the flefh at pleafure, and examine how much it was diffolved. In an hour the piece that lay in the ftomach was quite confumed, except a little cellular fubftance; but the other piece was entire. It was again introduced into the œfophagus, and re-examined an hour afterwards; but now the œfophageal liquor had begun to act upon the flefh: its weight at firft was fix pennyweights, but now only five and a half. It was kept upon the whole fix hours in the œfophagus, and loft nearly two pennyweights. Thefe experiments will not permit me to refufe to the œfophagus all power of digeftion, an effect undoubtedly produced by the fluid of the follicles (LXXV); but it is inconfiderable when compared with that of the ftomach, fince this vifcus diffolved fix pennyweights of flefh in an hour, while the œfophagus diffolved but two in fix hours.

LXXVIII. The experiments I afterwards made on young crows were ftill more decifive. The fame wire was employed, and two bits of flefh were faftened to it juft in the fame manner, the one lying in the ftomach, the other

other in the œfophagus. The former piece was generally quite diffolved before the folution of the other began, though in time this alfo was very fenfibly wafted. This diminution upon one occafion amounted to five pennyweights in the fpace of thirteen hours.

LXXIX. Laftly, in order to determine whether it was part only of the œfophagus or the whole of that canal in crows which poffeffes the power of folution, I formed a cylinder of flefh half an inch thick, and of the length of the œfophagus and ftomach taken together. I faftened this cylinder longitudinally to the wire employed before, and forced it down the throat of a crow, fo that one end touched the bottom of the ftomach, and the other reached almoft to the mouth. In a quarter of an hour the whole circumference of the cylinder was imbibed with a fluid, but at the lower end only, which refted upon the bottom of the ftomach had the flefh begun to be diffolved; here it was become whitifh. In about an hour for near an inch, i. e. for the whole length of the ftomach, fcarce any of the cylinder was left; and what little remained was gelatinous and had loft its cohefion, while the portion that lay in the œfophagus appeared to be unchanged: but it did not continue fo; a fort of erofion began to take place along the cylinder,

cylinder, and continued but with extreme flownefs. And as this erofion extended along the whole length of the cylinder, I had reafon to believe the whole length of the canal capable of concocting the food in a fmall degree, whenever it happened to be lodged there for feveral hours. But fuch an event never happens when crows take their food at pleafure, fince the pieces never exceed the length of the ftomach. In this refpect they differ from fome other animals, in which the food, after they have fwallowed it, reaches into the œfophagus.

LXXX. Upon confidering the great quantity of fluid continually dropping into the craw of gallinaceous fowls (LII), it appeared highly probable, that the concoction of the food is not a little promoted by the ftay it makes there before it falls into the ftomach. But the fact is juft the reverfe. The aliment is indeed foftened and macerated (LII), but I could never perceive that it was at all diffolved: at leaft, I could never fee any trace of folution on feveral vegetable matters, which had been long retained in the craw. In a fpace more or lefs fhort they become foft, and are imbibed with a fluid; but I have not been able to perceive, that they were in the leaft diffolved. We muft therefore conclude, that the

the œfophageal liquor in gallinaceous fowls is different from that of crows.

LXXXI. But why is the food fo foon digefted in the ftomach, and fo flowly in the œfophagus? Is it becaufe the gaftric fluid is more efficacious, or in greater quantity, than the œfophageal? What are the properties and characteriftic marks of thefe two liquors? May we hope that experiments out of the body will be as inftructive as thofe made in it? To enable myfelf to procure a large quantity of thefe fluids at pleafure, was the firft ftep towards the folution of thefe problems. And as fuch a quantity could not be eafily got by killing the birds, it became neceffary to invent a contrivance for obtaining it from them alive. To put bits of dry fpunge into the tubes, and leave them fome time in the œfophagus and ftomach, appeared to be the beft means of attaining this end; for they muft neceffarily be faturated with the liquor of thefe cavities, and when vomited or drawn up, will fupply the experimenter with a confiderable quantity, provided he ufes a number. Three tubes were introduced into the ftomach of a crow, and four hours afterwards vomited up; the three little fpunges, when they were taken out and preffed between the fingers, afforded thirty-feven grains of gaftric liquor,

liquor, which was frothy, of a turbid yellow colour, had an intermediate taſte between bitter and ſalt, and being ſet to ſtand in a watch-glaſs, depoſited in a few hours a copious ſediment. As the ſediment appeared to ariſe from the food that was diſſolved by the gaſtric juice (for the bird had taken food a little before it ſwallowed the tubes) I repeated the experiment upon another crow, of which the ſtomach was empty, and continued ſo till the tubes were thrown up. This precaution I ever afterwards obſerved, at the ſame time taking care, that the faſt did not laſt too long, leſt it ſhould induce a morbid ſtate in the animal. I was likewiſe careful to cleanſe the ſpunges from every impurity, by repeated waſhings and dryings, before I made uſe of them again. Upon repeating, with theſe precautions, the foregoing experiment with the three tubes, I obtained thirty-three grains of gaſtric juice in a ſtate of purity. It differed from the former in being of a tranſparent yellow colour, and in depoſiting very little ſediment; it had the ſame bitter and ſalt taſte. It appeared to have very little volatility, as it was kept ſeveral days in a watch glaſs without ſuffering almoſt any diminution. When thrown upon burning coals, it extinguiſhed them inſtead of taking fire; and when brought near a candle, it did not rife in flame. Further, paper foaked in it and thrown upon the fire, did not burn till the gaftric fluid was evaporated. Nor had it more volatility or inflammability when juft taken from the ftomach and ftill warm.

LXXXII. The quantity, which was not inconfiderable, obtained from three fpunges, gave me hopes of collecting enough for chemical experiments at large, and for attempting artificial digeftion. Every crow was capable, as I found upon trial, of taking eight inftead of three tubes; and as they would be thrown up in a few hours, I could repeat the experiment feveral times a day. Therefore to five crows, of which I then happened to be in poffeffion, I gave forty tubes furnifhed with little fpunges, i. e. eight to each. In three hours and a half all the tubes were returned by the mouth, and the quantity of gaftric fluid expreffed from them amounted to four hundred and eighty-one grains. In a few days before I had collected thirteen ounces of liquor. I employed it for the purpofes for which I had defigned it, and which fhall be mentioned in their proper place.

LXXXIII. While I was engaged in thefe experiments, I obferved feveral remarkable facts. The firft was, that the gaftric juice

flowed

flowed in great abundance into the cavity of the ſtomach; it ſometimes happened, that one of the ſpunges was brought up in a quarter of an hour after it had been ſwallowed, and in this ſhort ſpace it was conſiderably loaded, and in an hour as much as it could be. Secondly, when a conſiderable quantity of fluid has been obtained, another, and even a third may be got immediately. Sometimes when a crow has vomited up the eight tubes, I put freſh ſpunges into them and returned them without delay; and this I repeated a third time and found, that the quantity procured the laſt time was as great as the ſecond, and even the firſt. Thirdly, the fluid had always the qualities above-mentioned (LXXXI), if we except a difference of colour. It is commonly of a pale orange, but ſometimes of a cineritious yellow.

LXXXIV. I took the ſame method to procure the œſophageal liquor, with the variation only of a ſingle circumſtance. The tubes were now faſtened to threads, which were brought out at the mouth, and tied round the beak to prevent its being opened.

Thus the tubes were fixed in the œſophagus, without the leaſt danger of their getting into the ſtomach, or being thrown up. Beſides, I could draw them up at pleaſure. I
intro-

introduced four tubes at once into the œfophagus of a crow, and extracted them again in three hours. I learned from this firſt trial, the ſcantineſs of the œfophageal compared with the gaſtric fluid. The four ſpunges ſupplied me with eleven grains only. Doubting whether this might not be accidental, I repeated the experiment ſeveral times, and allowed the tubes to remain longer in the œfophagus, but the ſpunges were very far from being ſo thoroughly ſaturated with fluid as in the ſtomach: ſo that direct experiment proves the great abundance of the gaſtric, in compariſon with the œfophageal liquor. If the ſtomach and œfophagus of a crow be laid open longitudinally, the latter will be found to be moiſtened with its proper fluid only, while the former generally affords reception to part of it likewiſe. Theory too, in the preſent caſe, agrees with fact. The natural poſture of crows, and indeed of moſt other birds, is ſuch, that the liquor which oozes out from the internal ſurface of the œfophagus, muſt deſcend to the lower parts from the law of gravity, and thence into the ſtomach. This organ muſt therefore be the receptacle of the œfophageal fluid; but it is more than probable, that it has a peculiar fluid alſo (LXXVI): beſides, we are certain, that the

bile

bile is mixed in confiderable quantity with the gaftric juices. I have very frequently found the bottom of the ftomach in crows full of it, and this is the reafon why this juice is always bitter and yellow. Further, upon opening the duodenum longitudinally, I have perceived the yellowifh green veftiges of the bile, which is difcharged into that inteftine at the diftance of at leaft three inches from the pylorus through the cyftic duct, which evidently arifes from the gall-bladder. The conjunction of all thefe liquors, muft produce a quantity of fluid far larger than that which derives its fource from the œfophagus alone. And I doubt not but this is the reafon, why the food is digefted more fpeedily and perfectly in the ftomach than the œfophagus (LXXVII, LXXVIII). Though I fhould alfo fuppofe, that this is in part owing to the greater energy of the gaftric liquor from the admixture of the bile, which never rifes into the œfophagus, as appears from the juice of that canal never being at all yellow or bitter, but nearly infipid and colourlefs.

LXXXV. It remains now for me to relate fome attempts to produce artificial digeftion with the gaftric juices, referving for another opportunity the recital of the chemical experiments made upon that fluid, obtained both

from

from crows and other animals, with the view of acquiring as complete a knowledge as poſſible of its nature and properties. The great abundance I was able to procure from crows, by means of vomiting, gave me the advantage of inſtituting a greater number of trials, than with that of gallinaceous fowls (LVI, LVII), from which the gaſtric liquor could not eaſily be procured without killing them. I firſt wiſhed to examine the effect of the gaſtric juices of crows upon fleſh, in the open air. It was January, and Reaumur's thermometer placed near the veſſel uſed for the experiment, ſtood at the fourth and fifth degree (*a*). For greater certainty in theſe experiments, I eſtabliſhed a term of compariſon, by employing ſimilar veſſels containing the ſame fleſh infuſed in water. I alſo took care upon the preſent as well as other occaſions, that the fleſh ſhould be completely immerſed in the reſpective liquors, and that the phials ſhould be cloſed with ſtopples. For ſeven days the fleſh kept in the gaſtric juice, and in water continued the ſame. On the eighth I

(*a*) Wherever the thermometer is mentioned in this work, the ſame, viz. Reaumur's is to be underſtood.

N. B. The fourth and fifth degrees of Reaumur's thermometer anſwer to about forty-two and forty-three and one-fourth of Fahrenheit's.

perceived a flight folution, for upon agitating both liquors, feveral particles feparated from the larger mafs, and fell down to the bottom of the phials. No further progrefs was afterwards made, and the gaftric fluid did not feem at all more efficacious than common water; only the flefh immerfed in the former was preferved from putrefaction, but not in the latter.

LXXXVI. In this experiment I had ufed beef; I verified the fame obfervation upon the more tender flefh of calves, chickens, and pigeons, notwithftanding the heat of the atmofphere had now raifed the thermometer to feven degrees (*a*). While I was making thefe experiments in the natural temperature of the air, I was employed about others of a like nature in a warmer medium, viz. in a ftove, in which the heat varied between 22°(*b*) and temperate. And now the effects produced by the gaftric fluid, differed from thofe produced by water. In the latter the flefh began to be a little diffolved in two days; this was the effect of incipient putrefaction, as appeared plainly from the fœtid fmell which began to be exhaled. The fmell continued

(*a*) Forty-eight and three fourths of Fahrenheit's.
(*b*) Seventy-nine and a half of Fahrenheit's.

to

to increase during the following days, and in a week became intolerable, when the flesh was reduced to a nauseous pulp. In the gastric juice the solution was more rapid, and exhibited very different phænomena; twenty-five hours were sufficient to decompose the flesh contained in it, and in a little more than two days there remained only a very small morsel entire. These solutions never emitted any bad smell; whence it is evident, that they did not arise from incipient putrefaction, like those in water, but from a more efficacious and a different menstruum, viz. the gastric liquor.

LXXXVII. Being now engaged by different occupations, I was obliged to interrupt these experiments, and could not resume them till the June following; and then taking advantage of the season, I exposed to the sun two phials filled to a certain height with gastric juice from crows, in one of which were immersed several pieces of beef, and in the other crumb of wheaten bread. Nine hours of sunshine much forwarded the artificial digestion, which was the object of enquiry. A good part of the flesh was reduced to a kind of glue, that when it was handled adhered to the fingers; nothing like flesh remained in any of the pieces, but the nucleus, which was still

con-

consistent and fibrous, which two qualities it lost the next day; after having been exposed six hours to the sun the nuclei, like the outside, no longer retained a fibrous structure. In the sun, the heat as well on the first as the second day, was between forty and forty-five (*a*). The gastric liquor produced upon the bread a change analogous to that which the flesh had undergone. It not only lost its white colour and turned grey, but had become viscous, and no longer presented to the eye the appearance, though it retained somewhat of the taste of bread. Of bread as well as flesh immersed in water and exposed to the sun for the same time, there was a perceptible diminution; but it was very superficial and inconsiderable when compared with that produced by the gastric fluid. The bread turned sour, and the flesh became putrid, circumstances that did not take place in the least, in the other phials.

LXXXVIII. Thus a tolerably complete concoction was obtained in the heat of the sun; but it was reasonable to suppose, that it would be still more perfect in the temperature of the stomach. In the preceding dissertation I have

---

(*a*) An hundred twenty-two, and an hundred thirty-three and one-fourth of Fahrenheit's thermometer.

observed, that by way of substitute for the natural heat of the animal that furnished gastric liquor for the experiment, I fixed the tubes under my axilla (LVI, LVII); such an expedient was necessary in that case, since glass is incapable of resisting the violent action of the gizzard. But there was now no longer the same danger, and the experiment might be made in the following manner. Several glass tubes six lines long and three in diameter, were hermetically sealed at one end, and at the other bits of flesh were introduced, and then the tube was filled with gastric fluid. It was then very carefully stopped with sealing-wax, and the tubes were forced into the stomachs of several crows. Should digestion now take place it might be properly called artificial, since it must have been effected in close tubes, to which the juices of the stomach could have no access. But I soon found that the wax became soft in the animal heat, and consequently did not keep the tubes closely stopt, as I wished. There was however no difficulty in substituting a firmer cement, which would not be either melted or softened; and with such a cement I repeated the experiment just mentioned, and others of a like nature, which I shall describe hereafter. I prepared two tubes

tubes in this manner; they were given to a crow, and returned by vomiting in an hour and an half. I will not conceal my amazement at finding, that the pieces of flesh inclosed in the tubes were not in the least changed, unless it was in having acquired a blueish red colour. My amazement was still more increased upon observing, that they had undergone no further alteration after remaining four hours longer in the stomach of the same crow, enclosed as before in two sealed tubes. These bits of flesh weighed in all twenty-eight grains; so inconsiderable a quantity would have been dissolved in a few minutes, if it had been loose in the stomach, and in a very few hours, if it had been enclosed in tubes open at the ends.

LXXXIX. Did this unexpected failure arise from the communication between the external air and that within the tubes being cut off, or from a deficiency of gastric fluid, or else for want of the action of the stomach upon the flesh? I considered maturely these conjectural explanations, but they appeared altogether insufficient. With respect to the last, it is repugnant to all those facts which prove the solution of aliment within tubes, open indeed at the ends, and perforated along the sides, but which effectually prevent the mechanical action of the ftomach upon their contents. That the gaftric fluid was in too fmall quantity for the folution of the flefh, is a fufpicion unworthy of attention; for the pieces were always covered by it, fo that the quantity of fluid muft have been greater than that of folid. Laftly, the communication between the external air and that within the tubes being intercepted, cannot in all likelihood be the reafon why folution did not take place. In order to determine this certainly, I made the following curious experiment. Having prepared feveral glafs tubes of the length of fix inches, I fealed them hermetically at one end, by means of a reverberated flame, and the oppofite extremities were drawn out fo as to form elongated cones. Through the open end of thefe cones I poured a quantity of gaftric fluid, together with a few fmall pieces of flefh, which filled two thirds of the wider part of the cone. I then introduced the cones by their bafis into the ftomachs of fome crows, allotting one to each bird; and when they refted upon the bottom of the ftomach, their apexes came out at the mouth. To prevent their being thrown up, I ufed the precautions mentioned in another place (LXXVI). Thefe conical tubes muft no doubt have been very incommodious

modious to the animals, but they were exceedingly well adapted to the end I had in view, since a free passage was allowed for the external air into them. However, notwithstanding this, the flesh remained several hours immersed in the gastric fluid, without shewing any sign of decomposition.

XC. It is proper to apprize the reader, that when the sealed tubes, or the cones, were kept long in the stomach, as, for instance, ten or twelve hours, the flesh was generally reduced to a dark-coloured gelatinous pulp. But this did not remove my surprize at seeing so slow a solution in those close receivers, in comparison with the rapidity of the process in the stomach. The gastric juice was quite fresh, it was in sufficient plenty, and the flesh put in the tubes and cones was exposed to the same degree of heat when it is in immediate contact with the sides of the stomach.

If crows are killed during the process of digestion, the bottom of the stomach generally abounds in gastric juice, which when compared with that expressed from the spunges, appears to differ a little, being more dense and bitter, and of a yellow inclining to azure. The juice which is mixed with the food, and occupies the upper parts of the stomach, approaches more to the nature of that with which the spunges are imbibed. Having learned from experiment, that digestion proceeds most rapidly at the bottom of the stomach, on account probably of the gastric juice being more active and efficacious there from its immediate mixture with the bile, which gives it the yellowish azure hue and a bitterer taste, I preferred this juice to that from the spunges, and repeated with it the experiments with the sealed and conical tubes mentioned in the LXXXVIIIth and LXXXIXth paragraphs. But the event did not answer my expectation, no solution of the flesh taking place till several hours had elapsed.

XCI. Upon comparing the laboratory destined by nature for the process of digestion, and these receivers prepared by art to accomplish the same end, I could discover but two circumstances in which they differed; the flesh in the vessels undergoes the action of a fluid which is never renewed; while, on the contrary, in the natural laboratory it is continually subjected to the action of fresh juices, incessantly supplied by an innumerable multitude of follicular glands. Besides, the gastric juices being confined within the cavity of the stomach, there is little or no evaporation; whereas, when exposed to the air, and consequently fequently cooled, they cannot but lose some of their more volatile and active particles by evaporation. Does then the slow solution of flesh in close tubes and in cones, depend upon the gastric juice being deprived by these two causes of part of that energy, on which digestion depends? I found from experiment, that the former cause, at least, the want of renovation had great influence in retarding the solution. If, instead of perfectly closing the tubes, I left a small perforation capable of allowing ingress and egress to the gastric juice, the solution of the flesh took place much sooner. The same thing happened, when, instead of leaving the same juice in the cones all the while, I was at the pains of changing it several times. But warmth is another condition absolutely indispensable for rendering the gastric fluid of these animals fit for digestion. When this liquor is kept in a temperature not more than four or five degrees above the freezing point, its solvent power is so much impaired, that it does not seem more efficacious than common water (LXXXV). This is also observable at seven degrees (*a*) (LXXXVI). In order to render the effects of the gastric juice perfectly

(*a*) Forty-seven three-fourths of Fahrenheit's thermometer.

sensible,

sensible, a stronger heat is requisite, as from ten to twenty-two degrees (LXXXVI). Still solution proceeds very slowly; to remedy this the animal heat is necessary, viz. about thirty degrees (*a*) (XC). And so remarkable is the effect of heat in this particular, that the very liquor, which, for want of being renewed, dissolves flesh slowly at thirty degrees (XC), effects this very speedily at forty and forty-five degrees (*b*) (LXXXVII).

XCII. Every time I expressed the juice from the sponges, I washed them in pure water, which was tinged yellow by the remains. After having made so many experiments on the gastric fluid in a state of purity, I conceived it might not be altogether without its use, to make one with the water in which the sponges had been washed; with this water I therefore filled a small glass phial, which was left exposed to the sun, with a piece of flesh in it, for three days in July. The flesh (which was mutton) shewed some signs of solution. On the third day, there appeared upon the bottom of the phial a quantity of impalpable matter of a cineri-

---

(*a*) Ninety-five one-half of the same.
(*b*) One hundred and twenty-two, and one hundred and thirty-three one-fourth of Fahrenheit's thermometer.

tious

tious colour, confifting of particles feparated from the flefh immerfed in the liquor. Notwithftanding the feafon, as it ufually is in July, was very hot, it had acquired little or no fœtid fmell; whilft a fimilar piece of flefh, expofed to the fun in the fame manner, but immerfed in water, became intolerably putrid on the fecond day.

XCIII. But it is time to quit the fubject of digeftion in crows, and to proceed to that of herons, the other fpecies of birds which I propofed to examine in this differtation. The herons upon which my obfervations were made, and which the nomenclators denominate *cineritious*, or *grey* (*a*), muft certainly be claffed among birds with intermediate ftomachs, fince the fides of this vifcus have an intermediate thicknefs and folidity between membranous and mufcular ftomachs. When this organ is dilated, it appears about two inches wide, and as many long; its form approaches to that of a cylinder. When opened lengthwife, and obferved internally, it prefents the appearance of rugæ, of which fome run in a longitudinal, fome in a tranfverfe, and others in an irregular and oblique direction. The fides of the ftomach are co-

(*a*) Linn. Syft. Nat. T. 1. Bel. Av.

vered

vered with a kind of gelatinous lining, of some consistence, but easily removed, and of a colour between white and yellow. This lining seems organized; and I should be inclined to suppose, that it is the innermost coat of the stomach. The nervous coat next presents itself; it is of a whitish colour, and moderate thickness, but its texture is strong, and it is not easily lacerated. When this coat is cleaned and dried with a napkin, and then distended, and compressed underneath, it is immediately covered with an effusion of very small and scarce visible drops, which enlarging and approaching towards each other, form at least a thin aqueous covering. And if this be wiped away, and the nervous coat be again distended and squeezed, another like the first will appear; and in like manner a third, a fourth, &c. with this difference only, that the quantity of moisture is every time diminished. There can be no doubt of this being a portion of the gastric fluid, discharged directly into the cavity of the stomach. I have employed the utmost attention in searching whether this liquor derives its origin from glands, or any analogous bodies, but could never discover either the one or the other; and therefore we must suppose that it is secreted by small arteries, which open into the stomach,

ſtomach, and depoſit their contents there. After the nervous, we have the muſcular coat, of a red colour, and ſcarce a line in thickneſs. It is compoſed of fleſhy ſtriæ, partly tranſverſe, partly longitudinal. The former appeared to me to occupy the ſurface only, the latter conſtitute the internal ſtrata, and are continued to the termination of this coat. There is alſo another coat, confiſting of cellular ſubſtance, and this is the laſt of all.

XCIV. The ſtomach always, and eſpecially when empty, contains more or leſs gaſtric fluid, of a bitter taſte, turbid yellow colour, and generally of ſome denſity. The bitterneſs is owing to the bile, which has that taſte, but in an intenſer degree; I have often found it at the bottom of the ſtomach, and in the vicinity of the pylorus. The gall-bladder exceeds an inch in length; its greateſt diameter is of five or ſix lines; in ſhape it reſembles a ſmall egg, of which the ſharp end is inſerted into the liver. Notwithſtanding many careful examinations, I am not certain of having found the cyſtict duct; I however ſuſpect that it perforates the duodenum, at the diſtance of ſix inches from the pylorus; this I collect from a line of an azure-yellow hue, which ariſes from the gall-bladder, and is inſerted into that part of the inteſtine.

XCV. Above

XCV. Above the ftomach we meet with the fame kind of flefhy fafcia, which I have noticed in gallinaceous fowls and crows, (LXVI, XLVII, LXXV). In the grey heron it exceeds an inch in breadth. This fafcia is alfo covered with the fame gelatinous lining that invefts the ftomach (XCIII). Next we find a nervous coat of a finer texture than that of the ftomach, of which it appears to be a continuation. This coat, when attentively viewed, looks like a fieve, fo much is it perforated at every part. The perforations are nothing but the apertures or mouths of fubjacent follicular glands, occupying almoft all the infide of the fafcia, and vifible through it. If the nervous coat be any where compreffed, a vifcid, cloudy, and, as far as I could judge, infipid liquor oozes out at thefe pores, and continues to ooze out if the preffure be continued. The follicles that lie beneath, manifeftly fupply this fluid in fuch abundance. It would be fuperflous to defcribe thefe glandular bodies; fince they exactly refemble thofe of crows and gallinaceous fowls; whether we confider the immenfe number of them, or their pofition, contiguity, fhape, or colour. When we raife this aggregation of follicles, we come to the mufcular coat, which is very thin, and confifts

of

of several strata of long and compact fleshy fasciculi; next to which lies the last or external coat, the thinnest of all, and consisting of cellular substance.

XCVI. The œsophagus is about twelve inches in length, and in breadth one and a half. Its shape is nearly cylindrical, but it becomes narrower near the stomach. Upon examining it externally with a microscope, I discovered that it was quite full of minute bodies, which are, I suppose, glandular. When it has been carefully inverted and inflated, and the humidity, which always covers it, has been wiped away, if now we lay hold of one end, and squeeze it forcibly, so that it shall be enlarged in the adjacent parts, the humidity will appear again; and upon repeating the compression, it will be seen several successive times, just as in the stomach, (XCIII); with this difference, however, as I conceive, that the humidity in the stomach derives its source from small arteries, and in the œsophagus from minute glands, or some analogous bodies.

XCVII. It was natural to suppose that this apparatus of fluids, which are constantly trickling into the cavity of the stomach and œsophagus of herons, is chiefly designed for digestion. But the small number which I
possessed,

possessed, and their almost never vomiting, like crows, indigestible bodies, and consequently tubes, did not permit me to make such a series of experiments as I could have wished. I have however been able to make the most essential, of which one consisted in enquiring into the manner of digestion in the stomachs of these birds. For this purpose I had recourse to the tubes, than which I do not believe there are any means better adapted to such enquiries. It is well known that the grey heron feeds on fishes, frogs, water-snakes, and several sorts of aquatic worms and insects. Those in my possession devoured frogs, and especially fishes, with great greediness; and I therefore used them for my experiments. They swallow frogs of a moderate size whole. A whole frog, inclosed in a tin tube, was introduced into the stomach, together with a fish, of bulk nearly equal, included in another tube. In twenty-four hours the heron was killed, and the stomach opened; though the tubes were very thin, they had received no damage, if we except two slight contusions upon one of them; they were so light, that it was not difficult to guess that they no longer contained the same quantity of matter that had been put into them. The little fish was all dissolved, except some

of

of the ribs, a few bones of the head, and a bit of the flesh of the back, which had become so tender, that the parts no longer cohered. The frog's shape was more easily distinguishable than that of the fish; the pulp of the thighs, and even the bone itself, was quite destroyed; but the ends of the lower, as well as the upper limbs remained. The integuments of the abdomen and thorax were no longer to be found; and the subjacent flesh was become so soft, that it appeared to have undergone a slight boiling. The bones had acquired the softness of cartilage. These remains of the frog and fish were impregnated with gastric fluid, and tasted bitter. The intelligent reader perceives the immediate consequences of this experiment. In the first place the stomach of herons acts with some force upon the substances it contains, as we may collect from the slight contusions upon one of the tubes. Secondly, digestion, which was in an advanced state in the frog, and quite complete in the fish, is not the effect of trituration, but of the gastric fluid, which entering in at the open extremity of the tube, or through the holes at the sides, penetrated the two animals, and by virtue of its solvent power, partly consumed them; but it made greater havock upon the fish than the frog, on account of its being tenderer. Thirdly, the efficacy of the gaftric fluid of the heron is not limited to the folution of foft parts, fuch as the fkin, flefh, &c. but extends to the hardeft alfo, viz. the bones.

XCVIII. Of the laft circumftance I wifhed to obtain greater certainty, by putting bones alone into the tubes. We have already feen that crows are incapable of digefting hard bones, and that they digeft fuch as are tender with difficulty (LXX, LXXII, LXXIII). It was therefore an object of great curiofity to difcover what would happen in herons; and it was eafy to fatisfy this curiofity, by inclofing bones of feveral forts in the tubes; in one I therefore enclofed only the tender bones of frogs or fifhes; in another hard bones, viz. the thigh-bone of a turkey broken in two pieces. The pieces of both the hard and tender bones were formed into two bundles, and tied with thread. After an heron had retained thefe two tubes in its ftomach twenty-feven hours, it was killed. It was with a mixture of furprize and pleafure that I faw the tube which contained the fifhes and frog's bones empty, while the ftring remained entire. The gaftric juice had then perfectly diffolved the bones. But this was not the cafe with refpect to the contents of the other tube.

tube. I should have considered them as untouched, if they had not appeared smoother, whiter, and perhaps thinner than at first. They now weighed only eleven pennyweights and six grains, whereas at first they had weighed fourteen pennyweights; they had therefore lost of their original weight three pennyweights within six grains. If this experiment be compared with that made upon crows, with the same intention, 'it will appear evident, that the gastric juice of the latter is less efficacious in dissolving bones than that of the heron. And in truth, their nature requires that they should digest every part of the animals upon which they feed. I gave them some frogs, and observed their way of eating them; when of a moderate size they swallow them whole, when very large they separate them into several portions, and swallow them without parting the flesh from the bones. Since, then, herons do not enjoy the advantage of vomiting up substances incapable of being digested (LXXVII), and the bones of frogs and such animals cannot easily be voided at the anus, Nature has wisely endowed the stomach with the power of concocting and assimilating bone.

XCIX. It was equally curious and important to enquire, whether the œsophagus of the heron

heron as well as of crows, is capable of performing digestion (LXXVII, LXXVIII, LXXIX). The great length of the neck, and consequently of the œsophagus, is extremely favourable to such an enquiry. For this purpose a flead frog was forced half way down this canal with the head downwards, where it was fixed by a string, of which one end was tied round the hind legs, and the other came out of the mouth, and was wrapped round the neck. In this situation it remained two hours, and more was effected than I could expect in so short a space. The animal was indeed entire, but was become very tender, though the tenderness did not penetrate far below the surface. This appearance of incipient concoction induced me to push the experiment further, that I might see how it would terminate. The frog was therefore replaced in its former situation, where it continued three hours longer; I then thought it time to examine the animal, and drew up the pack-thread, but nothing came up with it except the hind legs and ribs; the rest remained in the throat, and an instant afterwards I perceived the animal transmit it into the stomach. I found the legs and thighs half dissolved, and being very desirous of knowing what had happened to the other

parts

parts of the animal, I determined to kill the heron without delay. The frog was in the ſtomach; the external muſcular fleſh was quite deſtroyed, and what remained entire was eaſily diviſible into ſeveral portions, eſpecially at the articulations. The ſame appearance of decompoſition had taken place, as if it had been macerated in water; but it did not exhibit the ſmalleſt ſign of putrefaction.

c. Although experiment thus abundantly evinced a ſenſible concoction in the œſophagus, I had not adverted to one circumſtance, which however deſerved attention; viz. to fix the preciſe loſs which the fleſh underwent in that cavity. I therefore repeated the experiment with this view, but having no frogs in my poſſeſſion, I ſubſtituted ſome fleſh with which I happened to be provided, conſiſting of cow's lungs, to the amount of half an ounce, forty grains. It was drawn out of the œſophagus by means of the pack-thread, after it had remained there thirteen hours, when it had loſt ſeven pennyweights and two grains.

As the œſophagus of the heron is membranous, it is more than probable, that its mechanical action did not concur in producing this effect; it however was proper to prove this by direct experiment, which might be done by means of the tubes. With them therefore I repeated the experiment which I had before made, in order to determine, whether the œsophagus of the heron is capable of digesting food. Solution of the flesh undoubtedly took place within the tubes, and so I was convinced, that it did not depend on any motion of the œsophagus, but on the efficacy alone of the fluid which is secreted by it.

CI. Another experiment yet remained to be made, whence we might deduce not only the exact diminution of the flesh, but the proportion between its diminution in the œsophagus and stomach; two globular pieces, two-thirds of an ounce each, of cow's lights were introduced, one into the œsophagus, and the other the stomach; each remained seven hours in its respective situation, when the heron was killed; the ball from the stomach at first of the size of a walnut, was now no larger than a pea, and weighed only twenty-eight grains. That which had lain in the œsophagus was indeed reduced in bulk, though but very little in comparison with the other; it had lost three pennyweights eighteen grains.

Both these instances gave me an opportunity of remarking, that the juices, whether

of

of the œsophagus or stomach, did not seem to act by penetrating deeply into the substance of the flesh, but by corroding the surface; the external layer was first dissolved, and then those that lay beneath in their order. And in reality, when I came to wash the ball taken from the œsophagus, and wipe away the external gelatinous stratum already dissolved by the œsophageal fluid, the next stratum shewed the natural fibrous, firm, and red appearance; and when the ball was cut into two hemispheres, the inside seemed perfectly found, without the smallest sign of having been impregnated or touched by that corrosive liquor. The same observation is applicable to the other ball; for notwithstanding its great diminution, it was quite found within.

I had now but two herons left, and these I sacrificed to the desire of ascertaining still further the excessively rapid concoction of the stomach, compared with that of the œsophagus. And in fact, I observed it again on two frogs and as many fishes, of which the former continued eight hours in the œsophagus and stomach, and the latter nine hours.

These experiments incontrovertibly prove, that the œsophagus of the heron as well as of the crow, has the privilege of digesting any food that may happen to be lodged in it: this privilege extends likewife to other animals, as we fhall fee in fome paffages of the following differtations.

CII. The obfervations related in the prefent and the preceding differtation, prefent us with various inftances of agreement and difagreement in the digeftion of birds with mufcular and of thofe with intermediate ftomachs. Let us here, for the convenience of the reader, collect into one point of view thefe fcattered traits; they may fix more firmly in the mind all that we have obferved, whether curious or interefting, in thefe two claffes of animals. With refpect to the traits of refemblance, they may all be reduced to the relation between the gaftric fluids. Firft then it has been proved, that thefe fluids, befides being alike in colour, are always falt and bitter; and that the bitter tafte derives its origin from the bile, which regurgitates through the pylorus into the cavity of the ftomach. Secondly, thefe fluids are the immediate agents of digeftion, both in mufcular and intermediate ftomachs, independently of trituration. Thirdly, In thefe two orders of birds the fluids act in the fame manner in the folution of the food; they firft foften and next convert the furface into jelly,

then

then produce the same effect upon the interior parts, and so insinuating themselves gradually till it is completely dissolved. Fourthly, they do not entirely lose their solvent efficacy as soon as they are taken out of the stomach, provided they are heated to a proper degree, as artificial digestion proves. Lastly, The sources from which these fluids spring, are, in great measure, the same in both classes, viz. the follicular glands, with which their organs abound.

CIII. The differences are in part reducible to the inferior efficacy of the gastric fluid in muscular to that of the same fluid in intermediate stomachs. Thus the gastric fluid in the former is incapable of dissolving the same aliment, which in the latter it easily dissolves. In like manner the food, which each kind of gastric juice decomposes and digests, is sooner subject to this change from that which belongs to intermediate stomachs. And this is also the reason, why artificial digestion succeeds much sooner with the first than the second. The same inefficacy that the gastric juices of birds with muscular stomachs shew in decomposing certain aliments of a firm texture, extends also to their œsophageal juices in decomposing soft substances, notwithstanding the latter are tolerably well decomposed by the œsophageal juice of birds with intermediate stomachs. The prodigious effects of trituration in muscular stomachs, constitute another very striking difference between these two classes of birds, the feeble force of intermediate stomachs being scarce comparable with the enormous power of the other kind. Such a degree of force was absolutely necessary in these, since the juices are incapable of decomposing food of considerable firmness, such as seeds, the natural food of birds provided with gizzards; and therefore an agent capable of breaking, triturating, and thus pre-disposing them for digestion became necessary; and such are in reality the gastric muscles in these fowls.

DISSERTATION III.

CONCERNING DIGESTION IN ANIMALS WITH MEMBRANOUS STOMACHS. FROGS. NEWTS. EARTH AND WATER-SNAKES. VIPERS. FISHES. SHEEP. THE OX. THE HORSE.

TO examine at full length the nature of digestion is the object of these dissertations. By extending my enquiries to the three classes to which all animated beings may be referred, I hope to be enabled to solve the problem in a satisfactory manner. Of these classes the first comprehends animals with muscular, the second with intermediate, and the third with membranous stomachs. The last class is infinitely more numerous than the two former. If we suffer our imagination to range over the immense multitudes of quadrupeds, fish, reptiles, birds of prey, not even excluding man himself, we shall

shall find, that they are all, or nearly all, endowed with membranous stomachs; not to mention that numberless tribe of minute beings, the greater part of insects. My task would have been endless if I had projected enquiries, I will not say concerning every species of animals included under these genera, a project which many academies would not be able to execute, much less a single observer; but concerning great part of them. I was therefore obliged to confine myself to researches upon a small number. These researches combined with others already related in the two first dissertations, will be sufficient, if I am not very much mistaken, to set the theory of digestion in a clear point of view, both in animals and man. As the various species, which I take into consideration, cannot be all exhausted in a single dissertation, I shall distribute them into several, beginning with the animals that are situated lowest in the scale of sentient beings, and ending with that which occupies the highest and noblest place, with man.

CV. Let us begin then with frogs and water newts, two species of small carnivorous animals. As the mouth and œsophagus in the former are large, it was easy to introduce tubes into their long stomachs. But I was
soon

soon aware, that it would be neceſſary to make experiments upon a great number at once, if I wiſhed to know what changes the fleſh encloſed within the tubes underwent in the courſe of ſeveral days; for the tubes were very often vomited up, and at uncertain intervals, ſometimes in a few, ſometimes in ſeveral hours after they had been ſwallowed; at others again, after a whole day, and in ſome inſtances, after a ſtill longer interval. As I knew, that this ſpecies of animal very greedily devours any ſort of fleſh that falls in its way, I did not think of ſelecting, but took what happened to come firſt to my hands, and this proved to be a piece of the ſmall inteſtine of a ſheep, which I divided into twelve portions, and encloſed them in as many tubes. Theſe tubes were diſtributed among ſix of my largeſt frogs, two being given to each. They were kept in a very large veſſel of water with high perpendicular ſides, that they might not make their eſcape. I neglected the tubes that were thrown up, and examined thoſe only which remained in the ſtomach. In the ſpace of a day I obſerved the following reſults. From the intervals of the grating which lay before the open extremities of the tubes, there oozed out a cineritious matter,

which,

which, when touched, adhered to the fingers, and formed long filaments. When the grating was removed, I perceived, that this gluten was nothing but the flesh itself, which at that part began to be decomposed and to change its nature, retaining however the characteristic marks of flesh in the more internal parts of the tubes. Upon opening the stomachs I did not find any gastric fluid; they were quite dry.

CVI. In two tubes that were examined at the expiration of two days, the flesh had undergone a further decomposition. It now not only oozed out at the meshes of the lattice-work, but likewise at most of the perforations in the sides of the tubes; and when it was drawn out with the point of a pair of forceps, and then freed by washing from the viscid mucilage, what remained of real flesh or intestine was so very little, that I do not believe it exceeded the thirtieth part of its original weight. At the end of the third day there remained but a single tube in one frog; in this there was no flesh, but it had been all dissolved to gluten, had oozed out of the tube, and adhered to the sides of the stomach, excepting a very small portion that was sticking to the tube. This viscid matter

was

was infipid to the tafte, a certain proof that the gaftric fluid had effected this alteration without the concurrence of any mechanical action of the ftomach. It muft however be allowed, that this fluid is exceedingly flow in producing its effects, fince it required three days. This flownefs muft have arifen either from the fmall quantity or inefficacy of the fluid, or perhaps from both caufes. In confequence of this tardy action I found in fimilar experiments upon fix other frogs, that the flefh in fome of the tubes was not entirely confumed at the end of the fifth day.

CVII. The gaftric liquor of frogs is not however on this account incapable of concocting in time fubftances which we fhould have fuppofed above its power, viz. bones. In a quantity of frogs brought me one day by the fifhermen, there was one fo large, that I was induced by its enormous fize to kill it, in order to fee what it contained; I found, that the enlarged bulk was owing to a moufe in the cavity of the ftomach. The hair had begun to fall off, and the fkin was become fo very tender, that it had loft its cohefion. The fore as well as the hind legs had undergone a greater degree of folution, the bare bones only being left, and they were confiderably
wafted

wasted and converted into a semi-gelatinous substance. The mouse upon being opened appeared quite sound internally, the destruction was confined to the surface, and therefore occasioned by the gastric fluid, which had begun to act here on the external parts, just as it does in animals with muscular and intermediate stomachs. The smallness of the extremities permitted the fluid to penetrate them with greater facility, hence it had almost consumed them without sparing even the bones. In this instance I could not perceive any sign of trituration, for the mouse was neither bruised nor lacerated; nor can I conceive what other force can be exerted by stomachs composed of such fine coats, besides that of compressing the bodies they contain, when they happen to be very large.

CVIII. The mouth and throat of water-newts are both so narrow, that they would not admit the usual tubes, they however admitted others made in the same form, but of a smaller size, on purpose for them. From having kept these animals in my house for several years, both when I had occasion to examine the circulation of the blood, and to observe the admirable reproduction of their limbs, I had learned, that the food which
they

they devour with the greatest avidity is living earth-worms (*a*). Nearly the same observation is made by my illustrious friend, Mr. Bonnet, in his fine memoir *concerning the reproduction of the limbs in water-newts*, in which he confirms my discovery of this wonderful reproduction in the clearest and most decisive manner, after it had been questioned by Messrs. Adanson and Bomare, for want of address and skill in making experiments in this branch of zoology (*b*). I had then recourse to earth-worms; they were cut in pieces, and placed alive in the tubes, which were introduced into the stomachs of several salamanders. The gastric fluid of these little reptiles acted more speedily than that of frogs (CVI). The divided worms began to change colour in fifteen hours, and to become soft and flabby. About the thirtieth hour the parts had lost their cohesion, and the

---

(*a*) I treat at full length of these aquatic lizards in my three works intituled,

Prodromo di un' opera da imprimersi sopra le riproduzioni animali,

Del Azione del Cuore ne' vasi sanguinei,

De' Fenomeni della Circulazione observata nel giro universale de' vasi.

(*b*) The memoir is inserted in Rozier's Journal for November, 1777.

N. B. It is likewise reprinted in the late collection of his works.

rings were no longer visible; and in less than two days they were converted into a whitish pulp, of which the greatest part had run out of the tubes.

cix. The dissection and examination of the stomachs of newts presented me with a phænomenon, which must not be concealed from the reader, both on account of its singularity, and the light it throws on the present subject. This phænomenon is nothing less than a great number of small white worms in this viscus, visible to the naked eye; of the thickness of a thread, and the length (at least in the largest) of two-thirds of an inch; however if we wish to examine them minutely, we must employ the microscope. They are of two sorts; in one both extremities terminate in a point, the other has one end pointed; but the other obtuse, and marked with a dark spot; the latter species is shorter than the former, and thinner in the same proportion. Each species is furnished with rings, narrower at the ends of the body, and wider at the middle, as is generally the case in anular worms. These two sorts of worms are not flat or compressed, but round; it is therefore certain that they do not belong to the genus of *tænia*, or the *gourd-worm*, but to that of round or columnar worms *(teretes)*. They are not loose in the cavity of the

the ſtomach, as the worms lodged in the inteſtines of larger animals commonly are, but are always found with one extremity inſerted to ſome depth in the internal coat of that organ; hence it requires ſome force to detach them, and frequently they break ſooner than ſeparate. Of thoſe that have the dark ſpot, the more obtuſe end is fixed in the ſtomach; it is impoſſible to ſay whether this is the caſe with reſpect to the others, ſince both ends are equally pointed. The looſe extremity projects into the cavity of the ſtomach, ſometimes coiled up in the form of a circle, and at others twiſted in a ſpiral. If the ſtomach be taken out of the animal, and ſet to macerate in water, the worms live for many hours without quitting their ſituation: if afterwards we ſeparate them with the hand, without breaking them, and place them upon ſome ſubſtance, in order to obſerve their movements, they will be ſeen to writhe in various directions; ſometimes bringing the mouth towards the tail; ſometimes ſtretching themſelves in a right line; and at others making ſtrange contortions, as is uſual with reptiles in general.

CX. Not being able to conjecture for what purpoſe the part perpetually inſerted in the ſubſtance of the viſcus, could be deſigned,
unleſs

unless it was to suck the thinnest and purest part of the liquor; and consequently supposing it to be the head of the animal, or at least some analogous part, I tried to discover the mouth with the aid of the microscope; but my endeavours were vain. I believe, however, that I found the alimentary canal; it is a bright silver-coloured species of intestine, running along the worm, in a tortuous manner, from side to side; it is always full of a number of particles, which fluctuate regularly, like a buoy, probably impelled by a sort of periftaltic and antiperiftaltic motion. This canal is common to each species; in that with a dark spot at one extremity (CIX), a second canal may be perceived; it is strait, and probably (I should rather say certainly) the receptacle for the eggs; for I have always observed it more or less full of a great number of corpuscles, of an oval shape, floating in a very transparent lymph; these corpuscles, when the worm is not in motion, always continue at rest. If we lay hold of the animal by its extremities, and break it in the middle, the little canal will generally be broken, and the ovula will make their escape in a stream from the lacerated part. It is not difficult to burst them between two pieces of talc, when a thin fluid sprits from them;

after

after which the eggs become dry and opake, confisting now of nothing but the empty cover, as always happens to the membranaceous eggs of small animals. Each worm of this species is furnished with those oval particles enclosed in their canal; if they are real eggs, as there is great reason to believe, we must conclude that every individual is an hermaphrodite; it will however remain doubtful whether they are strictly so; i. e. have no need of copulation, like sweet-water polypes, and many other sorts of microscopical animals, &c. or else, in the wider acceptation of the term, are like testaceous and naked snails and earth-worms; each of which brings forth eggs and living young, but requires the concurrence of another individual.

CXI. I might probably be asked whether these worms lodge in healthy newts, or rather in such only as are diseased. This doubt suggested itself to me; and in order to clear it up, I examined not only such as I kept at home, and were therefore liable to the suspicion of unhealthiness, but such also as were newly caught, and full of health and vigour; but the stomachs both of the one and the other harboured alike these unpleasant guests. But it must be observed that they do not fix their abode in all newts; and that in those

Vol. I.   K   where

where they do, they are not equally numerous. Of the immenſe number I have opened at different times, and with different views, three-fourths have had a family of worms in their ſtomachs; which is ſometimes compoſed of only five or ſix individuals; at others of ſeveral dozens, and at others again of an hundred or more.

CXII. In my numerous examinations of the ſtomachs of the different animals mentioned in this work, crows alone have exhibited a phænomenon nearly reſembling what is found in newts; I mean a quantity of worms lodged in the ſtomach. But theſe worms are not inſerted into the internal coat, as in newts, but are found between the internal and the nervous. We are very well acquainted with the little worms that live in trees, and generally fix their abode between the bark and the wood; and lurking there unſeen, devour the cortical part, which furniſhes them with an agreeable aliment. If the bark ſhould be parted from the trunk on purpoſe, or by accident, their devaſtations are expoſed to view, in the form of excavated paths, winding backwards and forwards in a ſerpentine direction; nor is it uncommon to ſurprize the worms actually employed in forming theſe excavations, which ſerve them at once for food

food and lodging. The fame thing nearly is obfervable with refpect to the worms of crows. If the internal be parted from the nervous coat flowly and carefully, thefe animals are fuddenly expofed to the eye, adhering for the moft part to the back of the internal coat, lurking in certain cavities formed in its fubftance, and which in all likelihood, arife from the erofion of thefe very worms. Further, we find fome with both ends expofed, while the middle is deeply buried in the fubftance of the internal coat. Laftly, others have one extremity inferted into this, and the other into the adjacent nervous coat; but they never make their way into the cavity of the ftomach. Thefe worms do not appear to differ from thofe in newts, in colour, length, thicknefs, or in the alimentary canal; they have however one effential difference, they are without rings, but have a fmooth and flippery fkin. In their motions they are dull and languid; when taken from their abode, and placed in water, they live many hours. They are found both in grey crows and rooks; but I have never feen them in any part of the body except the ftomach.

CXIII. But let us return to the worms in newts (CIX, CX, CXI), and confider them in as far as they relate to digeftion. I affert that

that their prefence is an incontrovertible proof, that no fenfible degree of force is exerted by the ftomach; for how is it poffible to conceive that the fides of the ftomach can rub againft each other, or at leaft impinge againft the food, without doing the fmalleft injury to worms of fo delicate ftructure? I have more than once taken the ftomach of a newt between my thumb and finger, and compreffed it very gently, or rubbed it lightly, and upon opening it, have always found fome rupture, fome difcontinuation of the parts of thefe worms. We muft therefore conclude, that digeftion in water-newts is folely the effect of the gaftric fluid, of which the efficacy has been already fhewn in the decompofition of earth-worms inclofed in tubes (CVIII). I have alfo feen its action, in a manner equally ftriking, on worms which newts have taken and fwallowed of their own accord. How tenacious thefe minute reptiles are of life, is abundantly proved by cutting them into feveral pieces, in confequence of which they do not die; but on the contrary multiply, as many worms being produced as parts into which they were divided (*a*). It is true, they do not ceafe to live after having remained

(*a*) See Reaumur, Bonnet, and my Profpectus.

ten or twelve hours in the ſtomach of a newt; nay, when they fill it too full, they void ſome alive and crawling by the mouth, whether by actual vomiting, or whether the worms, after various movements in ſo diſagreeable a place of confinement, at laſt find their way out through the œſophagus. But they certainly die at laſt, not becauſe they are triturated or cruſhed to pieces, for they continue whole for ſeveral hours; the gaſtric fluid firſt ſoftens and then converts them into a gelatinous ſubſtance, and by a continuance of its action, at length reduces them to an impalpable maſs.

CXIV. But how come the tender worms in the ſtomach to eſcape ſolution, when all other inſects, whether aquatic or terreſtrial, upon which the newt feeds, die and are digeſted? If it ſhould be ſaid, that this happens becauſe they have been habituated to the ſtomach by long reſidence, the difficulty would be perhaps removed to a greater diſtance, but certainly not taken away altogether. As the cauſe of this phænomenon, we muſt aſſign the inability of the gaſtric fluid to decompoſe theſe minute beings, however powerful may be its energy upon others of a ſtructure leſs delicate; juſt as a chemical menſtruum is capable of diſſolving one metal, but not another. Thus aqua regia diſſolves gold,

gold, but not silver; or an acid that combines with the calcareous, has no attraction for the argillaceous and filiceous earths. Nearly the fame difference of digeftion is alfo obferved in that of polypes provided with arms; they fometimes fwallow their own arms along with infects; but though the former die and are digefted, the fecond do not in the leaft fuffer. Thus a polype inferted into the ftomach of another polype, continues to live as before (*a*).

cxv. But let us proceed to ferpents, of which I propofed to treat after frogs and newts. Thofe which are moft eafily procured in the environs of Pavia are certain terreftrial fnakes, called in fome provinces of Italy, *Smiroldi* (*b*); and water-fnakes, which many naturalifts call fwimming (natrices) (*c*). The firft confiderably exceed the natrices and vipers in fize. The largeft are about an inch and a half in thicknefs towards the middle of the body, and forty-five and fometimes fifty inches long. The lower part of the body is white mixed with yellow and green ftreaks,

---

(*a*) Trembley, Mem. fur les polypes.
(*b*) Not defcribed by Linnæus or any other naturalift, as far as I know.
(*c*) *Natrix*. Linn. Syft. Nat. T. 1. Natrix torquata. Ray. quadr.

the

the upper part is blackifh, but towards the neck and head interfperfed with a milky white. They fly with greater fpeed than the water-fnakes, and far greater than vipers. They are not inferior to the latter in a fpirit of revenge, and their bite alfo draws blood, as I have myfelf experienced, but is harmlefs. Before I made ufe of the tubes, I wifhed to acquire fome knowledge of the œfophagus and ftomach. Having therefore fkinned one, and blown up the œfophagus in fuch a manner that the air could neither pafs out above nor at the pylorus, it appeared to me to refemble a large inteftine, cylindrical for about the length of nine inches, and becoming gradually narrower below, fo as to form a funnel of the length of four inches and an half. I foon perceived, that this funnel was the true ftomach, and the inteftine the œfophagus. Both the trachea and lungs run along the œfophagus, to which they are firmly attached by means of a membrane, as alfo is the heart, which has the fhape of an elongated pyramid, fituated at the origin of the lungs. We find likewife a vifcus arifing from the bafis of the heart, afcending upwards along the œfophagus, and adhering in great meafure to the trachea: it is of the fame length as the lungs, but its fubftance is different, being tender

and

and afh-coloured; I could not then determine what it was. Next below the lungs lies the liver, which, together with the *vena portarum*, refembles a long narrow leaf attached to a very long footftalk; both adhere to the œfophagus. Below the ftomach we find the fpleen, nine lines in length and of a very acute oval fhape. The gall-bladder lies in the region of the fmall inteftines, confequently at a great diftance from the liver; when we prefs it the duct is filled with bile, which it evidently difcharges into the duodenum at about the diftance of an inch from the pylorus. Near the gall-bladder we find another body fmaller than it, attached to the duodenum, and of a flefhy confiftence. I fhould fuppofe it to be the pancreas.

CXVI. If we feparate the œfophagus and ftomach from the lungs and other parts juft defcribed, and open it longitudinally, the œfophagus appears fimply membranous; the membrane of which it confifts is very thin and of a filver colour. The ftomach is compofed of thicker fides, and among the coats which compofe it we have one of flefh, which like the flefhy coats of other membranous ftomachs, is very thin. I could not perceive, that the œfophagus is provided with any glands or follicles; but I obferved, that the
ftomach

ſtomach was abundantly ſupplied with them throughout its whole length; they diſcharge part of their liquor on being preſſed, and the internal coat is moiſtened with it.

CXVII. I come now to experiments relative to digeſtion. I found great facility, not only in paſſing the tubes into the ſtomach, but likewiſe in bringing them up whenever I pleaſed. I made an aſſiſtant lay faſt hold of the ſnake ſo as to prevent its ſtriking or wreathing round the body, while I opened the mouth and forced a tube in lengthwiſe, and then, by means of a thin rod, thruſt it two or three inches down the throat. After this the reſt followed of courſe; for I had only to preſs with my fore-finger and thumb the neck of the animal in the place oppoſite to the top of the tube, which was forced to deſcend for ſome way down the œſophagus, and by a repetition of the ſame manœuvre I ſoon brought the tube to the bottom of the ſtomach, which I knew by the reſiſtance it made when I attempted to puſh it lower; for now the narrow paſſage of the pylorus prevented its deſcent. By a like preſſure, but made in the oppoſite direction, from below upwards, I could bring up the tube from the ſtomach into the œſophagus, and thence out through the mouth. I employed this contrivance

trivance for introducing the tubes into the ſtomach, and bringing them out at the mouth in water-ſnakes likewiſe, and even vipers, managing the laſt however with ſome care, which is very requiſite, in order to avoid being bit by theſe ſerpents during the operation, when they are highly exaſperated.

CXVIII. When I was opening ſome of my land-ſnakes *(ſmiroldi)* to examine the alimentary canal, I found in the ſtomach of one a wall-lizard not in the leaſt injured or digeſted. I thought of employing it for my experiments, as it muſt be a kind of food well adapted to theſe reptiles. I therefore encloſed a piece of the tail of this lizard in a tube, which continued for a whole day in the ſtomach without having its contents at all diſſolved. Thirty-ſix hours produced ſomething more. The tail of the lizard is compoſed of a number of little muſcles, enchaſed one within the other, and bound round by a thin anular membrane. The piece of tail was placed in the tube in ſuch a manner, that the inveſting membrane was in contact with the ſides, and the muſcles were bare at the open ends. The membrane had ſuſtained no injury, but the muſcles were eroded on the plane of ſection, and a little excavated. Upon touching them I found, that they had been

been converted into a gluten of fome vifcidity. The gaftric fluid then (for the mechanical action of the ftomach could produce no effect within the tubes, were any fuch action to be exerted at all) had begun to digeft the flefh, by diffolving what lay at the ends of the tube before it attacked that which was contiguous to the fides; not only becaufe it was not covered by the membrane, but alfo becaufe it had freer accefs at the ends: the folution however went on, though very flowly; for after the tube had been five days in the ftomach, a little of the mufcular flefh remained, and the membrane was almoft entire.

CXIX. The flefh of a lizard's tail is rather tough, and it was probable, that this circumftance had retarded the progrefs of digeftion; it was therefore proper to employ fome of a lefs firm texture; accordingly part of the liver of the fame animal was enclofed in the tube, and given to a fnake *(fmiroldo)*. In this inftance digeftion was more fpeedy; for in three days and an half the tube was quite empty.

But what if inftead of enclofing the flefh in tubes, we fhould introduce it into the ftomach without any covering? It was obvious to fuppofe, that it would be fooner digefted, fince the gaftric juice would have freer fcope

for

for its action. And so in reality it happened. A piece of lizard's tail of the same size as in a preceding experiment (CXVIII), did not require quite two days for its digestion; and a portion of liver, equal to that before-mentioned (CXVIII), underwent the same process in two and thirty hours. Of this I assured myself by opening the stomachs of the two snakes, one of which had taken part of the liver, and the other of the tail.

CXX. We come now to the water-snakes or the *natrices*. Nothing can be more striking than the resemblance between the stomach and the œsophagus in this, and the foregoing species. Besides the trachea, lungs, heart, liver, vena portarum, having nearly the same configuration, and lying on the same parts of the œsophagus; this cavity is very capacious and long, consists in like manner of thin membranous coats, and ends in a funnel, which is the true stomach of the animal. The gall-bladder too is about an inch distant from the lungs, and deposits its contents in the duodenum, by means of the cystic duct. The stomach also, as we have observed in the land-snake *(smiroldo)*, is furnished with a great number of follicular glands.

CXXI. It is easy to learn the nature of the food of water-snakes, and we ought in consequence

sequence to provide it for our experiments. Among the antients Oliger Jacobeus, where he treats of frogs, and among the moderns Valisneri will satisfy us, that these reptiles live chiefly upon frogs. Next to man water-snakes may be denominated their greatest scourge. They particularly frequent the water of ditches, puddles, ponds, lakes, such in short as is frequented by frogs; and here they make an easy prey of them, notwithstanding they mutually give each other notice when they perceive the snake at a distance, by a kind of whistle or outcy of distress, as I have often observed, at which all fly with the utmost precipitation: Dante was not acquainted with this circumstance.

> Come le rane innanzi l'inimica
> Biscia per l'acqua si dileguan tutte,
> Finchè alla terra ciascuna s'abbica (a).

A fisherman having brought me three very large and vigorous water-snakes, I gave each at the same time a tube enclosing a different part of a frog; one muscle, the other liver, and the third spleen. The tubes were left three days and an half in the stomach. Upon forcing them out, I observed the same kind

(a) Infern. Cant. 9. Fol. 161, &c. As frogs scour along the water, at the approach of the water-snake, without stopping, till they have gained the dry ground.

of digeſtion that I had before ſeen in frogs (cv, cvi). The fleſh was beginning to be changed into an adheſive cineritious gluten; the interior parts were unaltered. The tubes were now introduced a ſecond time into the ſtomachs, and when they had continued there two days they were found empty; ſome of the adheſive matter ſtuck to the outſides of two of them.

cxxii. It is not unknown to naturaliſts, that this ſpecies of ſnake has no teeth, and is conſequently obliged to ſwallow its prey whole. In ſummer I have often taken them with whole frogs in the ſtomach. It was therefore not unreaſonable to ſuppoſe, that they are capable of digeſting the bones; and the leſs ſo, as it ſeems difficult for them to be voided backwards, on account of the narrowneſs of the inteſtines. It might indeed be ſuſpected, that theſe bones are vomited, as I have found to be the caſe with the tubes, both in this and the former ſpecies; but this is not a conſtant and regular evacuation, as in crows (lix) and birds of prey, as we ſhall ſee hereafter; but takes place at uncertain intervals, and ſometimes does not happen at all for ſeveral days. In order however to aſcertain the fact, I broke two tibiæ weighing nine grains each to pieces, encloſed them in the

tubes,

tubes, and forced them into the ſtomachs of two water-ſnakes. After they had continued there two days, they were become ſoft, and had loſt three grains. In five days more they were ſtill ſofter, and now weighed together only five grains. Soon afterwards the two ſnakes died, and it was not in my power, though I wiſhed it very much, to proſecute this curious experiment as far as it would go. From the beginning however of the progreſs we may ſuppoſe, that the bones would have been totally diſſolved, and conſequently it is highly probable, that water-ſnakes digeſt the bones of thoſe animals upon which they feed.

CXXIII. By the activity of the gaſtric fluid of the water-ſnake in digeſting not only fleſh but bone, I was induced to try to procure a quantity that I might examine it more particularly. For this purpoſe I employed ſpunges as before (LXXXI, LXXXII), and my ſucceſs exceeded my expectations. Six little ſpunges, that had lain two hours in the ſtomachs of three ſnakes, encloſed in tubes, afforded me enough to fill a watch-glaſs of a moderate ſize. It had the following qualities; the colour approached that of ſoot, it had the fluidity of water, and evaporated very ſlowly: it has both a ſalt and bitter taſte, and is not

inflam-

inflammable. Hence it appears to bear a very ftrong refemblance to the gaftric fluid of the other animals, upon which my experiments were made: this refemblance extends likewife to the odour, which is exactly like that of the fame juices in birds of prey, of which I fhall fpeak particularly in the next Differtation. I referve the account of fome chemical experiments made upon this fluid, till I fhall have an opportunity of fpeaking of the examination of the other gaftric juices which I have already mentioned, or fhall have occafion to mention in the prefent work.

CXXIV. We have before feen the ftrong analogy between the configuration of the ftomach and œfophagus in land and water-fnakes. In vipers thefe cavities have the fame general form; nor do they differ with refpect to the efficient caufe of digeftion. I repeated upon them moft of the experiments defcribed above: feveral tubes, furnifhed with different forts of flefh, were left in their ftomachs for a fpace more or lefs long, and the effect was juft the fame as in water and land-fnakes; it would therefore be fuperfluous to defcribe them particularly. It will be better to turn the reader's attention to fome experiments on thefe three fpecies of reptiles differently

ferently modified, but relative to the same subject.

cxxv. Having frequently opened these animals when newly taken, I have sometimes observed, that their stomachs are not large enough to contain the whole prey, and that part lies in the œsophagus. This part never shewed any mark of concoction, notwithstanding what lay in the cavity of the stomach was sometimes half digested. Thus, for instance, I have found five or six large beetles in the body of a land-snake or viper; those that lay in the stomach were scarce distinguishable, while, on the contrary, those in the œsophagus were entire, or nearly so. I once saw a frog with the lower limbs, which projected out of the stomach, not at all damaged, while the rest of the body lay in the stomach and was half reduced to a pulp. These experiments made by the serpents themselves gave me reason to suppose, that what takes place in them is exactly contrary to what happens in crows and herons, for the reader will remember, that in these birds the œsophagus is really capable of digestion (LXXVII, LXXVIII, LXXIX, XCIX, C); but in the animals in question it seems to belong exclusively to the stomach. A very simple experiment was sufficient to ascertain the

the point. Into the ſtomach of one of theſe ſerpents a frog, for inſtance, might be ſo introduced that part ſhould lie in the œſophagus. The frog might be faſtened to a cylinder of wood, and thus firmly fixed in the ſame place. The cylinder ſhould touch the bottom of the ſtomach with its lower extremity, and reach ſome way above that organ. I applied this apparatus to a water-ſnake, and at the end of the ſixth day opened it longitudinally. Upon examination my ſuſpicion that the œſophagus was without efficacy, was changed into firm perſuaſion. The lower limbs, the part of the animal that had lain in the ſtomach, had nothing left but the bare bones, whereas the whole body which had extended into the œſophagus had ſuffered no injury.

CXXVI. The experiments related in the CXVIIth and following paragraphs were made in April, when the animals had lately quitted their ſubterraneous lurking places, and ſtill retained ſomewhat of that torpor which benumbs them during winter. At this time digeſtion, as we have ſeen, is a very ſlow proceſs. Are we to preſume, that when they become more lively, active, and vigourous, as the heat of the ſeaſon increaſes, they likewiſe perform digeſtion more ſpeedily? for the

effect

effect of heat in promoting the operation of the gastric fluid appears from other facts (LXXXVII). This idea was suggested by re-perusing the fine memoirs of the illustrious Trembley on polypes, from which the influence of the temperature of the atmosphere upon the digestion of these wonderful animals is evident; insomuch, that the very food which in a hot season is completely digested in twelve hours, when it is cold requires sometimes two or three days. In order to determine whether the same thing takes place in my reptiles, I chose July for a term of comparison, when the difference, if any existed, must needs be more striking, as the thermometer in the shade stood at 22° and 23° (*a*); whereas in April, when the first experiments were made, it did not rise above the twelfth or fourteenth deg. (*b*). And now upon repeating the experiments already described, I found that heat has some power in accelerating digestion, but not so much as I had supposed. Flesh inclosed in tubes did not require above two days to be completely digested; and when an equal quantity was in-

---

(*a*) Eighty-one and an half, and eighty-three and three-fourths, F.

(*b*) Fifty-nine, and sixty-three and an half, F.

troduced into the ftomach by itfelf, about half that time was fufficient.

CXXVII. Naturalifts were already apprized of the tardinefs of digeftion in ferpents. In Bomare we read an account of a ferpent at Martinico, which retained a chicken in its ftomach for three months, and did not completely digeft it, for it ftill preferved fome traces of its fhape, and the feathers ftill adhered to the fkin (*a*). It is a circumftance deferving of particular notice, and which I fhall have occafion to apply in another place, that flefh does not become fœtid from remaining long in the ftomachs of thefe cold animals, as I have obferved in the courfe of my experiments, and efpecially in a viper, which having been kept above two months in my houfe, could not but be unhealthy; this individual retained in its ftomach for fixteen days a lizard, which had been previoufly macerated in the gaftric fluid; nor could I perceive that it had any odour, except that of this juice. And yet fuch was the heat of the feafon, that another lizard, about the fame fize, which I had placed out of curiofity in a clofe veffel, containing a little water, emitted an infupportable ftench before the expiration of the third day.

(*a*) Dict. d'Hift. Nat.

CXXVIII.

CXXVIII. But what can be the caufe of this flownefs of digeftion in ferpents? As they are cold animals, that is to fay, as their blood very little exceeds the temperature of the air, it may feem probable that this phæ-nomenon might be owing to the want of that heat, which is peculiar to animals of warm blood. And I fhould not have been unwilling to admit this caufe, if other animals, with blood equally cold, had not enjoyed the privilege of digefting their food in a much fhorter period, as we fhall foon fee (CXXXIV). We cannot affign a deficiency of gaftric fluid as the reafon, for their ftomachs abound with it (CXXIII). I cannot attribute it to any thing but the inefficacy of the fluid itfelf; and this is by no means fingular, for we have already difcovered a circumftance nearly fimilar in animals with mufcular ftomachs, in which the gaftric juices do not fo foon digeft the food, as in animals with intermediate ftomachs (CIII).

CXXIX. Of fifhes I fhall firft treat of that fpecies which bears fo ftrong a refemblance to ferpents, and is even confidered in the chain of animated beings, as the intermediate link between fifhes and ferpents, I mean the eel. The ftomach in this animal varies from the ftructure generally obferved by nature; it
is

is not a canal immediately continued with the duodenum, but a kind of blind gut, of confiderable length, ending in a point; after the food has been received into this gut, and been digefted, it muft afcend, and return to the upper part of the ftomach, in order to pafs into the duodenum, which forms an acute angle with that upper part. The natural figure of both may be feen in Blafius's Anatomy of Animals (*a*).

Into the ftomachs of four eels I introduced feveral tubes, containing pieces of fifh, the food moft agreeable to eels. In order to preferve them alive, I turned them into a fmall ftew, whence I could take them at pleafure. They were killed at the end of three days and eighteen hours; and the tubes were found at the bottom of the ftomach, entirely covered with a dark-coloured mucus, which, on attentive examination, appeared to be the remains of the fifh, that by this time was digefted. Upon wiping the tubes, and examining the infide, five out of eight were empty, and the three others contained a bit of flefh of the fize of a vetch, but it had loft its cohefion.

CXXX. This experiment abundantly proves, that in this fifh digeftion is produced by the

(*a*) Plate LII. Fig. 1.

gaftric

gastric fluid; I therefore proceeded to experiments upon such as are more justly entitled to the appellation of *fishes*. I chose for this purpose carp, barbels, and pikes, as they were the most easy to be procured. It has been long well known that the alimentary canal in many scaly fishes, is provided with one or more blind appendixes, which, because they lie in the vicinity of the pylorus, have been named *pylorici*; they are always full of a white, mucilaginous, and saltish fluid, which is discharged into the canal, and derives its origin from a number of glands lying in the appendixes. In some species they are few, in others in considerable numbers, and in others again exceedingly numerous; they amount in the sturgeon to an hundred; in those species, in which they are most numerous, the several fasciculi meet in a common duct; hence, notwithstanding their numbers, they discharge their contents into the pylorus (*a*) by a few mouths. In the three species I have just mentioned, this singular apparatus is not to be found; but the inside of the stomach and intestines is furnished with yellow bodies, that probably contribute in some way or other to digestion, though I have

(*a*) Haller, El. Phyf. T. 6.

not been able to afcertain their precife ufe. At firft fight they look like anular worms, adhering, as in newts, to the internal furface of the ftomach (cix); but if we lay hold of them with the forceps, the anular form vanifhes, and we find that they are real appendiculi to the ftomach and inteftines. When ftretched out, they are three lines long; each adheres to the villous coat by a footftalk. If we ftretch them till they break, a confiderable quantity of yellow liquor iffues out, and the body becomes more fhrivelled; and if it be now removed from the place of its infertion, we find under it a little tumour, through which a globule is feen indiftinctly; and if the tumour be cautioufly raifed, appears diftinctly: it is of a yellowifh white colour, from the liquor which it contains. Are thefe globules clufters of glands, and the vermiform bodies elongated ducts for conveying the liquor into the cavity of the ftomach? I would very willingly have adopted this notion, if I had not found that thefe fubftances, when compreffed from the lower end upwards, never difcharge their contents, either from the fummit, or any other part, contrary to what happens when we fqueeze the follicular glands in birds with mufcular, intermediate, or membranous ftomachs. On

this

this account I fufpend my opinion on the fubject; although I fhould incline to fuppofe that they are of fome ufe in digeftion.

CXXXI. Immediately under the teeth, the very beginning of the œfophagus in the carp is moiftened with a confiderable quantity of a whitifh turbid liquor, of a vifcid confiftence, and infipid tafte, which when wiped off is inftantly reproduced: and we here find a number of white papillæ, broad at their bafis, and terminating in a point, which, when preffed, emit the fame kind of fluid. If we make a gentle preffure any where near thefe papillæ, a fluid iffues out, but, as I fhould fuppofe, of a different nature, fince it is tranfparent, thinner than the former, and not at all vifcid. With the œfophagus, which is very fhort, and of confiderable thicknefs, is continued the ftomach, of a membranous ftructure, and very thin. It is eafy to diftinguifh two coats in this organ, the internal and the nervous; in the latter are buried thofe globules, which left me in doubt whether I ought to confider them as clufters of glands (CXXX). In this fhort defcription, we fee fources capable of fupplying the ftomach with a large quantity of fluid, notwithftanding it wants the *pylorical* appendixes.

CXXXII.

CXXXII. The conformation of the ftomach in the barbel, does not correfpond either with that of the carp, or various other fifhes. The œfophagus, ftomach, and inteftines conftitute a fingle gut, nearly as in earth-worms, and a variety of infects; this gut is only a little dilated at the ftomach, and contracted at the commencement of the inteftines. I could not difcover within this canal any veftige of glands or analogous bodies. However, both the œfophagus and ftomach are continually moiftened with a fluid in great abundance; which when we prefs or dilate either of thefe cavities, is feen to tranfude from the internal furface; and fince, according to every appearance, it does not arife from glands, we muft fuppofe that it comes from the open extremities of fmall arteries which terminate here.

CXXXIII. The ftomach of the pike has the fhape of a bag or fack, of much greater length than breadth; it is full of longitudinal rugæ, of a light flefh colour, and compofed of coats fo thin, as to be femi-tranfparent. The rugæ extend upwards into the œfophagus, which is eafily diftinguifhed from the ftomach by its white colour, and greater thicknefs. There is no appearance of glands either in the one or the other, though both, and especially the stomach, abound with liquor.

CXXXIV. As fishes are subject to vomiting, the tubes which I had introduced into the stomachs of my carps, barbels, and pikes, were frequently returned; and I was often chagrined at finding them, after a few hours continuance in the body, at the bottom of the vessel used for keeping them alive. However, from frequently repeating my experiments, though so many tubes were ejected prematurely, a few remained several hours in the stomach; and these were sufficient to satisfy my wishes. In the present case, the same thing happened which I had so often observed in other animals; the flesh was digested within the tubes, and that in a much shorter space than in serpents (CXXVI, CXXVII). This observation was verified on the barbels, carp, and pike; the two latter species exhibited a phænomenon, too closely connected with the present subject to be omitted. Happening one day to open a pike, I found within it a little fish, about three inches in length, lying longitudinally along the stomach, so that the whole head occupied the œsophagus. I had here a clear view of the origin and progress of digestion. The jaws of the small fish retained their natural colour, and appeared
unal-

unaltered. The eye was beginning to quit the orbit, and the gills had loſt their purple hue, and were become very ſoft. In the ſtomach the marks of digeſtion were more evident. The fleſh of the body was more and more tender as I proceeded downwards; and towards the bottom it had degenerated into a ſoft and ſhapeleſs maſs. The extremity of the tail, which had lain at the bottom of the ſtomach, was entirely conſumed, and with it the vertebræ of the ſpine, and the adjacent bones.

cxxxv. I met with another ſimilar circumſtance in a little carp. It had ſwallowed a ſmall lamprey, which was ſtretched out at full length, and occupied the whole ſtomach, and at leaſt two thirds of the œſophagus. The part that lay at the bottom of the ſtomach was changed into a kind of mucilage, in which there was no appearance of any thing organized, except ſome of the dorſal vertebræ. The parts that lay higher ſtill cohered, but they came away from the animal on being touched. The others, which occupied the œſophagus, ſhewed likewiſe marks of an incipient concoction.

Nothing can be more inſtructive than theſe two facts combined. They ſhew, in the firſt place, that the bottom of the ſtomach digeſts

more

more quickly than the parts fituated above, as we have feen in other animals (XC): fecondly, that the œfophagus, as well as the ftomach, is in fome meafure capable of concoction, a circumftance that has been already noticed with refpect to crows and herons (LXXVII, XCIX, C, CI); and which phyfiologifts have obferved before me in other fifhes. Laftly, that digeftion in the œfophagus is flower at its beginning, and in its progrefs; two things that have been remarked in the birds juft mentioned.

With refpect to the triturating power of the ftomach in thefe three fpecies of fifhes, not to mention that digeftion has been obtained in the tubes without its concurrence, I am of opinion that it has no exiftence; this I infer, from no effect being produced by it upon the tubes, upon which I have never perceived the fmalleft bruife, contufion, or injury, in my experiments on fifhes, any more than on frogs, newts, and ferpents, though they were fo thin, that the flighteft force would have been more than fufficient to diftort or bruife them.

CXXXVI. From cold animals let us proceed to fome experiments on the ftomachs of warm animals, fuch as fheep, oxen and horfes.

Reau-

Reaumur, in his second and last Memoir (*a*) concerning digestion, after relating at length his observations on a kite, slightly touches on some experiments upon dogs and sheep. I will here quote the results of his experiments on the latter species, reserving those on the two others for another place. Desirous of seeing whether digestion in sheep is the effect of the gastric fluid, he forced down the throat of one of these animals four tin tubes, two of which were full of fresh blades of grass, and the two others of chopped hay. Fourteen hours afterwards the sheep was killed and opened, when the four tubes were found in the first stomach, with their contents; and the grass and hay were not in the smallest degree digested, and but little softened.

Suspecting that they would undergo further alteration, and be even digested by a longer continuance in the stomach, Reaumur caused eight other tubes to be prepared in the same manner, i. e. four to be filled with fresh grass, and the remaining four with hay. The grass before it was put into two of these tubes, and the hay before it was put into two others, were moistened with human saliva. All the eight were forced down the throat of a sheep, which was killed thirty hours afterwards;

(*a*) Hist. de l'Acad. Roy. An. 1752.

during

during this interval, the animal had been kept ſtrictly faſting; this precaution had alſo been obſerved with reſpect to the former ſheep, that had not retained the tubes ſo long. In the courſe of the thirty hours, the greater part of the tubes were voided at the anus, but a few remained in the firſt ſtomach.

But neither had the graſs or hay undergone the ſmalleſt degree of digeſtion; they preſerved their original form and dimenſions; and when they were pulled at the two oppoſite extremities, reſiſted efforts to break them with the ſame force that ſimilar pieces of graſs or hay, that had been a little macerated, would have done. Hence it is inferred by this illuſtrious naturaliſt, that digeſtion cannot be effected in the ſtomachs of ſheep by a ſolvent, unleſs that ſolvent be aided by trituration: he was however ingenuous enough to confeſs, that theſe two experiments are of themſelves very far from throwing ſuch light upon the preſent ſubject as he could have wiſhed.

CXXXVII. The firſt thing I undertook with reſpect to ſheep, was to repeat exactly Reaumur's experiments. Thinking the tubes I had hitherto employed too ſmall, I had ſome made eight lines in length, and four in diameter. But I could not at firſt introduce them

them into the ftomach. After I had put them into the throat with my hand, though I pufhed them as far as my fingers would reach, they were always returned. I was unacquainted with Reaumur's method, for he does not give the leaft information about it. At laft an expedient occurred; it confifted in putting the tubes, provided with their contents, into a hollow cane, and introducing this cane into the œfophagus; I could now pufh them forwards with a rod, till they dropped out at the lower end of the cane into the œfophagus; and as the part of the œfophagus into which the tubes were now introduced, lay at a great diftance from the mouth, the animal, in fpite of all the efforts he made to return them, was obliged to receive them into the ftomach: to the fame contrivance I had recourfe likewife for oxen and horfes. Six tubes were given to a fheep; in twenty-feven hours it was killed and opened; it had eaten nothing during all the time it retained the tubes; and this precaution was ftrictly obferved upon every fheep upon which experiments were made. Notwithftanding fo long a faft, the firft ftomach contained a large quantity of grafs, fomewhat triturated; and though it had fed upon this before the experiment, it was not yet digefted. In the
midft

midst of this grass, that was thoroughly imbibed with a greenish fluid, with which great part of the stomach was filled, lay five of the tubes; the sixth had passed to the second stomach, which may be considered as an appendix to the first. The herbs which I enclosed in the tubes, after impregnating with my saliva, were beet, trefoil and lettuce; for three tubes I had used them green, and for the rest dry. Upon opening all the six, I could not perceive that either the fresh or the dried plants had suffered any diminution, or undergone any degree of real concoction; they had only become a little tenderer, and the fresh herbs had lost their green colour; in short, the result of this experiment was exactly like that of Reaumur's.

CXXXVIII. I should then have supposed, that in these animals digestion depends on the triturating power of the stomach, if it had not occurred, that, as the herbs enclosed in the tubes had not passed further than the first stomach, they might not perhaps have felt the influence of that kind of gastric fluid, which is requisite for the concoction of the food: for it is very possible that this fluid may reside in some of the other stomachs, and especially in the fourth, in which the aliments of animals with four stomachs, such as sheep, are always found in the state of a very soft paste. Reaumur indeed did not perceive any sign of digestion even in the tubes that had been voided at the anus, and consequently must have passed through the other stomachs. But this observation rested upon a single experiment; and to illustrate still further a matter of such importance, it could not but be proper to repeat it. I therefore treated another sheep in the same manner, and allowed it to live thirty-seven hours afterwards, that the tubes might have time to pass beyond the first stomachs. They did in fact pass beyond them; and I found all six in the fourth, which answered the end I had in view; the three species of plants however, mentioned above (CXXXVII), both the green and the dry, were entire, and seemed only a little softened by maceration.

CXXXIX. I was now about to declare in favour of the necessity of trituration in this animal, when a doubt occurred. Neither Reaumur nor myself had adverted to a circumstance, which ever precedes digestion, both in sheep, and every other quadruped endowed with four stomachs, as goats, oxen, deer, &c. I mean rumination. We are taught, both by dissection and daily experience, that the food, when it has arrived at the second stomach, does

does not immediately proceed to the third, and thence to the fourth, but, on the contrary, returns, and re-afcends up the œfophagus; and when it has reached the cavity of the mouth, is mafticated and ground over again, and impregnated with a large quantity of faliva; this procefs is repeated, till it becomes fit to be digefted. I therefore entertained many doubts whether the phænomena obferved by Reaumur and myfelf, were not rather owing to the want of rumination than of trituration. Wherefore, in order to decide with certainty concerning digeftion in fheep, I perceived it would be neceffary to repeat the experiments with plants previoufly triturated. And I did not conceive that this trituration was any thing fo peculiar in ruminating animals, that it could not be fupplied by man, provided he mafticated the herbs well, and impregnated them thoroughly with faliva. I therefore performed this eafy operation; the ufual tubes were employed; in three of them pieces of the green plants were enclofed, and in three others of the dried; but both had been well mafticated: the lines and nerves that traverfe the leaves, were however eafily diftinguifhable. Left, when thus broken down and divided, they fhould pafs out at the lateral pores, or through the mefhes of the lattice-work, I thought it would be proper to enclose each tube in a linen bag; supposing that in the present case that it would not be broken, since that muscular action, which is so considerable in gallinaceous fowls, does not exist in the animals in question. I gave the six tubes to a ram, together with six others, filled with the same plants, but not previously masticated, that I might be able to form a comparison. Fourteen hours after the animal had taken them, he vomited three at once; and in thirty-three hours five more were voided at the anus; at the end of two days he was killed. Of the four remaining tubes, two were found in the stomach, and the other two at the end of the duodenum; the bag in which they were enclosed was entire. The tubes that had been vomited had received more or less injury; the contents of two had not been masticated, nor had they undergone the smallest alteration. The contents of the third had been masticated, and were evidently wasted; for they now occupied little more than half the tube, whereas at first they had entirely filled it; they had acquired a subacid taste. Many of the pieces having lost their natural firmness, broke when I attempted to stretch them; the nerves only made some resistance.

In

In two of the tubes voided at the anus, the pieces of plant had not been mafticated; thefe did not feem at all diminifhed, nor was the cohefion of the parts deftroyed; on the contrary, the plants in the other three (which had been mafticated) were reduced almoft to nothing; the fmall remains confifted of bare nerves, with a little of the leaf attached; and both the one and the other were fo much foftened, that the flighteft force was fufficient to break them. The bag in which they were enclofed was dyed green, particularly in its infide; when twifted and preffed between the fingers, it yielded a livid yellow juice, of an acid tafte. This was far from being the cafe in the bags which contained the two tubes, of which the contents had not been chewed; their infide had fcarce a fhade of green; and this fhade was ftill lefs perceptible in the juice expreffed from them. With refpect to the tubes found in the laft ftomach, and at the end of the duodenum, the contents of the former had acquired a deep green colour, were a little macerated, but had not loft much of their natural firmnefs, and did not appear to be diminifhed in bulk. They had not been mafticated, but thofe of the two others had; and of them there remained only fome of the largeft nerves, which were themfelves very tender,

tender, and half decompofed. I have already obferved, that the tubes voided at the mouth were more or lefs bruifed; but all the reſt were quite free from injury.

CXL. The reader is already aware of the immediate confequences of thefe experiments. In the firſt place it appears, that the gaftric fluid of ſheep has no effect in digefting plants, unlefs they have been previoufly mafticated; otherwife it can only produce a flight maceration, nearly as common water would do in a degree of heat fomewhat exceeding the medium temperature of the atmofphere. Secondly, this fluid is abundantly capable of digefting plants, provided they are previoufly reduced to pieces by maftication; its firſt effect is to foften them, and deſtroy their natural confiftency; it then proceeds to diffolve them, not even fparing the tougheft parts, fuch as the nerves of the leaves; of this folution we have a clear proof in the green colour that appears on the linen enclofing the tubes, and in the juice expreffed from it (*a*). Thirdly, the triturating power of the ſtomach does not

at

(*a*) During my agreeable refidence at Geneva in the fummer of 1779, I had what I had long wifhed for, the fatisfaction of being perfonally acquainted with my illuftrious friend Mr. Bonnet, and of enjoying much of his converfation. I had alfo an

oppor-

at all contribute to digestion in sheep, but this process is entirely effected by the gastric juices. Fourthly, no such power exists in sheep, as we see from the tubes that were voided at the anus and found in the stomach having sustained no injury, notwithstanding pressure alone between

opportunity of taking his opinion on some productions, which I designed to publish, and particularly the present work concerning digestion. Three other reputable philosophers and excellent judges of the subject were present at the reading, Mr. Abraham Trembley, Mr. Johannes his worthy nephew, and Mr. Senebier, Librarian of the Republic of Geneva; and it did not appear to me, that my labours were disapproved by this respectable assembly. Mr. Bonnet gave me a book to peruse on the same subject, which, as it was new to me, made us apprehensive, lest the author should have anticipated me; it was intituled, "Essai sur la Digestion, & sur les principales Causes de la Vigueur & de la Durée de la Vie. Par Mr. Batigne, M. D. Berlin, 1768, 12mo." But I was soon aware, that Mr. Batigne and myself had pursued very different paths; in his book he does not enter into any experimental enquiry concerning digestion, but confines himself to reflections, which, although they are very pertinent and sensible, are calculated rather to excite than satisfy the reader's curiosity. Hence I should not have mentioned it, but for some objections started against Reaumur's Memoirs on Digestion. These I shall touch upon in a few short notes, at such places of the text as they seem most connected with. And here it is proper to mention one objection relative to the digestion of ruminating animals, which, before I was acquainted with Mr. Batigne's book, I had myself urged against Reaumur, and which experiment proves to be perfectly just. It consists in shewing, that the French naturalist had omitted the mastication of the plants enclosed in the tubes before he introduced them into the stomachs of sheep, which was the reason why they were not digested.

Nearly

tween the fingers is sufficient to flatten them. The contusion of the tubes that were vomited is no proof of the contrary, since it is evident, that this contusion was produced by the teeth of the animal during rumination. Lastly, these vegetables acquire a slight acidity during solution, but of this we shall have another opportunity of speaking hereafter (*a*).

CXLI. This quadruped feeds not only upon grass, but upon corn also whenever it meets with it; it is likewise very fond of bread. In order therefore to confirm still farther what has been advanced, I thought it would be proper to make an experiment upon some kind of grain. I selected wheat for this purpose; and as wheat may be procured under the various forms of seed, flour, and bread, I chose to make trial of all three. Six tubes were filled, three with those substances without any other preparation, and three others with the same after they had been well mas-

Nearly the same objection is urged by the learned physician in these terms; " The experiments (of Mr. de Reaumur) upon ruminating animals are still less conclusive; the grass contained in the tubes could be only macerated, since it was neither chewed nor broken down a second time by rumination." (L. c. Troisieme Reflexion sur les Experiences de Mr. de Reaumur). It was a piece of justice due to Mr. Batigne not to overlook this passage.

(*a*) In the last dissertation.

ticated.

ticated. The tubes were enclosed in linen bags as before, and given to a lamb six months old. It was killed thirty hours afterwards; none of the tubes were either vomited or voided at the anus; they were found partly in the third and partly in the fourth stomachs. The result of this experiment coincided with that related above (CXL). The grain, flour, and bread, that had not been masticated, were indeed penetrated thoroughly by the gastric fluid, but not at all dissolved. On the contrary, the corn which I had first bruised with a pestle and then ground between my teeth and reduced to a coarse paste, was in great measure consumed; nothing remained in the tubes but fragments of the bran, with some small remains of farinaceous matter adhering. The like had happened to the flour and bread, what remained consisted of a mucilaginous mass, without any appearance of what it had originally been. This matter had a slight degree of acidity, a quality which was far more evident in the bread, flour, and grain that were not dissolved by the gastric fluid for want of previous maceration.

CXLII. The vast quantity of gastric fluid with which ruminating animals are continually supplied, was already known to physiologists, and particularly to the great Haller.

After

After a faſt of two whole days, I have found thirty-ſeven ounces in the two firſt ſtomachs of a ſheep. It was green, but I know not whether this colour is natural to it, as the yellow hue is to that of crows (LXXXI); or rather whether it is adventitious, and derives its origin from the plants on which theſe animals feed, of which, notwithſtanding ſo long a faſt, there were ſtill ſome remains in the two ſtomachs. The great quantity of juice I had collected induced me to try whether, like that of ſeveral other animals, it was capable of digeſting food out of the body. I therefore encloſed ſeveral pieces of leaves of lettuce in two ſhort glaſs tubes (which I had previouſly filled with the juice), and ſealed them with wax at each end. The contents of one tube, as before, were maſticated, while thoſe of the other were left untouched. It was proper on the preſent occaſion to employ a term of compariſon, by repeating the ſame experiment upon two other tubes filled with water. That theſe four tubes might be expoſed to a degree of heat nearly equal to the temperature of ſheep, I fixed them under my axillæ, two under each axillæ, where they continued forty-five hours. The leaves immerſed in the gaſtric fluid, which had been previouſly macerated, had undergone

gone no inconsiderable change. Besides the loss of their bright green colour, they were converted into a kind of glue, in which it was just possible to find, with the point of a penknife, a few nerves, which were the only remains of the organization of the plant. This was far from being the case with the leaves that had not been masticated; for all the pieces were distinguishable, and the only difference was, that they did not make so much resistance as at first. The leaves immersed in water, both those which had been chewed, and those which had not, had not lost either their colour or consistence. From this comparison it appears, that the gastric fluid does not act on the plant as a mere aqueous fluid, but as a real solvent, nearly as it acts in the stomach itself. Nor was the heat to which it was exposed under my axillæ a condition without its part in the production of this incipient digestion; for in pieces of the same leaves of lettuce masticated in the same manner, but kept in my apartment, of which the temperature was about sixteen (*a*) deg. there appeared only a superficial maceration, notwithstanding they remained immersed in the same gastric fluid for the same space of time.

(*a*) Sixty-six, Fahren.

CXLIII. I closed my enquiries concerning the digestion of ruminating animals by some experiments on oxen. In these the same tubes and plants were employed as before, and the results perfectly coincided with those obtained from sheep; only in the present instance, Nature was more speedy in her operation. In less than twenty-four hours the tubes, which had been given to two oxen, were voided along with the excrements not in the least contused or injured. When taken out of the linen bags and examined, they were found to contain little more than the bare ribs and nerves of the leaves of beet, lettuce, and trefoil (which leaves had been previously masticated). The nerves were also in some degree macerated, and the slightest force was sufficient to break them. On the contrary, pieces of the same plants that had not been subjected to maceration were indeed slightly concocted, and their colour was a little faded, but they were entire. When applied to the tongue, they tasted subacid, like those which had been in the stomachs of sheep (CXXXIX, CXLI).

The horse does not chew the cud, but he resembles the ox in the membranous structure of his stomach, and the food upon which he lives. I was therefore desirous of seeing what changes masticated plants would undergo go by continuing a certain time in the stomach of this quadruped also, enclosed as usual in tubes. Here too they were digested, as I learned from some lettuce and trefoil enclosed in two tubes, which were voided in fifty-two hours.

CXLIV. When I reflect upon the various animals to which my enquiries concerning digestion have been hitherto extended, I perceive, that the ruminating species very nearly resemble birds endowed with muscular stomachs, with respect to the action of the gastric fluid. In both, that fluid requires an agent capable of breaking down and triturating the food, before it can dissolve and digest it. From the mouth of granivorous birds, where it undergoes no real alteration, the aliment passes immediately into the craw, where it is softened and macerated; from this receptacle it descends into the stomach: the triturating power of this organ performs the office of teeth, and breaks, grinds, and, if I may so speak, pulverizes it, and thus renders it fit to be dissolved by the gastric fluid, and converted into chyme. Nature employs a similar contrivance in ruminating animals. The hay and grass descend immediately into the first and second stomachs, in nearly the same state as when they were browsed. Here they are softened by the great

quantity

quantity of gastric juices, as feeds in the craw of birds with gizzards. But as the stomachs of ruminating quadrupeds have no sensible triturating power (CXXXIX, CXL, CXLIII), and the food requires trituration, nature has wisely provided for this by causing it to ascend, in consequence of a gentle stimulus to vomit, into the cavity of the mouth, where, by means of rumination, it receives the necessary predisposition to be digested by the gastric fluid, as happens to the food in the stomachs of granivorous fowls, after they have been properly triturated by the gastric muscles.

DISSERTATION IV.

THE SUBJECT OF DIGESTION IN ANIMALS WITH MEMBRANOUS STOMACHS CONTINUED. THE LITTLE OWL. THE SCREECH OWL. THE FALCON. THE EAGLE.

CXLV. REAUMUR having treated in his firſt memoir of the mode of digeſtion in granivorous and herbivorous fowls which are provided with gizzards, proceeds in his ſecond to enquire into the nature of that function in carnivorous birds, of which the ſtomach is membranous. From the facts related in the firſt memoir he concludes, that there does not exiſt in the gizzard any ſolvent capable of ſeparating the particles of the food. This ſeparation, he thinks, is effected by a force reſembling that exerted by mill-ſtones,

viz.

viz. the action of ftomachs of this conftruction upon their contents. He is moreover of opinion, that the facts adduced in the fecond memoir prove the exiftence of a menftruum in membranous ftomachs, capable of diffolving and digefting the aliment without borrowing any aid from the action of the folid parts.

As the great object of the firft differtation was to enquire by experiment into the mode of digeftion in fowls with mufcular ftomachs, I had there an opportunity of confidering fully Reaumur's experiments on that fubject; and we have accordingly feen, that the confequences he has deduced from them are by no means to be admitted in their full extent. This is plain from the XXXIX, XL, XLI, XLII, XLIII, XLVth paragraphs, to which, for the fake of avoiding ufelefs repetitions, I refer the reader. The prefent differtation, in which the fubject of digeftion in membranous ftomachs is continued, is the proper place for confidering the experiments related in the fecond memoir. As of all fowls birds of prey approach neareft to man in the ftructure of the ftomach, he chofe one of thofe large kites that are common in France, for the fubject of his enquiries. The periodical vomiting, common to all birds of prey, allowed the French naturalift to make a variety of experiments on

the

the same individual. He employed tin tubes filled with different substances, especially flesh, which, after having been some time in the stomach, were thrown up, and gave him an opportunity of examining the effects produced upon the contents. That the flesh was more or less digested according to the length of its continuance in the body of the animal, was the general and invariable result observed by Reaumur (*a*). Hence he justly infers, that in this case digestion is produced by the gastric fluid, without the concurrence of any triturating power. He mentions some

(*a*) Mr. Batigne thinks, that flesh enclosed in tubes was insufficient to convey a precise idea of the alteration it undergoes in the stomach, as it is only macerated in tubes, and not digested. " On voit de plus que la viande mise dans les tubes ne peut donner une idée precise des changemens qu'elle subit dans l'estomac de l'animal, puis qu'elle n'y est que macerée & non point digerée." *(L. c. premiere Reflection sur les experiences de M. de Reaumur.)* The author must allow me to observe, that in this attack he misrepresents Reaumur, who, p. 465, &c. of the Mem. of the Roy. Acad. expressly says, that the flesh given to the kite was not merely macerated or softened, but completely digested, and at last entirely consumed. He might indeed have objected to the small number of his experiments, as insufficient to ascertain the efficient cause of digestion, if that philosopher, whose ingenuousness was equal to his skill, had not perceived and publicly owned it himself. That tubes, provided the experiments are properly made and varied, are well adapted to shew the change produced upon food in the stomach, will be abundantly proved by the facts adduced in this treatise.

Vol. I.   N   other

other experiments, which I shall have occasion to consider below, and concludes from analogy, that digestion in other birds with membranous stomachs is produced in the same manner. He laments, however, that from the death of his kite, and his neglecting to substitute other animals in its stead, he could not adduce facts sufficiently numerous to illustrate the subject fully. He promises to supply the deficiency on some future occasion, but his death, by which a few years afterwards natural philosophy lost one of its great ornaments, prevented him from fulfilling his promises.

CXLVI. I do not presume that I shall be able to accomplish, what neither this illustrious naturalist, nor any other, as far as I know, has effected. But simply with the view of continuing my observations and reflections on digestion in fowls with membranous stomachs, I shall relate some experiments on different birds of prey, of which some seek it by night, and others by day. Among the former, I have used such as I could most easily procure, the little owl and the screech-owl. The food which I gave the first-mentioned species (*a*),

---

(*a*) This species is called by Buffon *petite chouette*, Hist. Nat. des Oiseaux, T. 2. Ed. in 8vo. and by Linnæus *strix passerina*, l. c.

and

and which it eagerly devoured, has enabled me to folve, among other problems, one that exercifed the fagacity of M. de Reaumur. Finding that the gaftric fluid of the kite digefted flefh, he wifhed to know whether it would alfo digeft vegetables; a circumftance he did not think probable, when he confidered the repugnance carnivorous birds fhew for them; and fo in fact it happened. When beans, peafe, wheat, inclofed in tubes, had lain fome time in the ftomach, they were thrown up juft in the fame ftate as they had been fwallowed: nor did boiling difpofe them to be diffolved any better by the gaftric fluid. Some fparrows, which I gave my owls, afforded me an opportunity of obferving the fame phænomenon. As they fwallowed them whole, they of courfe would receive into the ftomach feathers and food not yet digefted by the fparrows, and confifting of grain or bread. Now, after the flefh has been digefted, the feathers are vomited generally in the form of a hard ball; and along with the feathers the grain, which, though it is much foftened by maceration, yet continues whole. And if the matted feathers be difentangled, we may generally perceive evident traces of bread. Hence we have a clear proof that the gaftric fluid produces no change on fuch vegetables.

CXLVII. This fact, simple as it is, shews two things, of some importance: first, that the stomach of this bird is really membranous, and without any power of trituration: this appears from the grains (CXLVI) continuing whole, though they had been soaked till they were become so tender, as to burst on being gently squeezed between the fingers. I would not however affirm that the stomach has no action at all; for the globular mats of feathers can only be produced by this viscus contracting as the flesh is digested.

The digestion of the bones also deserves attention. It cannot be said that they are voided along with the excrements; for I must soon have been aware of this, as I kept my owls in cages; nor, for the same reason, could it have escaped my notice if they had been vomited. I have indeed sometimes found two or three little bones, as a dorsal vertebra, or a piece of the cranium, among the matted feathers, but never any thing like the whole skeleton. We must therefore conclude that they are digested.

CXLVIII. Reaumur's kite was capable of digesting bone, though of the hardest texture, and enclosed in tubes (*a*). Though the experiment just related is sufficiently de-

(*a*) Mem. cit.

cisive,

cifive, yet, as the bones were loose in the stomach, in order to be absolutely certain that the effect was produced by the gastric fluid alone, it was proper to repeat it with a tube; with this view a piece of the thigh of a pigeon was put into one of the same tubes that I used before: thus two experiments, one on the digestion of flesh, and another on that of bone, were made at once. By long practice upon birds of prey, I learned how to keep the tubes in the stomach as long as I pleased. When I had given one of my tubes to an owl, after it had been full fed, I found it was not thrown up till all the food was digested. This observation is applicable to all other birds of prey. The same thing also happened when they were fed sparingly. All the difference was, that as the full stomach requires more time to be emptied, the tubes were retained longer, and *vice versa*, when they were fasting, the tubes were sure to be returned in two or three hours. This observation, together with the knowledge I had acquired from experience, of the time these birds take to digest a given quantity of food, enabled me to guess pretty exactly how long the tubes would continue in the stomach.

I return now to the tube in which part of a pigeon's thigh had been put. After seventeen

teen hours continuance in the ftomach, the bone was no where changed except at the broken ends, which were a little foftened. The flefh, by which it was covered, as well as the integuments, had begun to be diffolved, for the furface was become exceedingly tender. In fourteen hours more greater effects were produced. The flefh was confiderably wafted; the bone was fhortened at the ends, and was fo foft, as to yield to the preffure of the finger. In twenty-feven hours more, there was no remains of flefh or periofteum, and the bone was a good deal fhorter than at firft. I could not but be defirous of feeing the end of the experiment, and therefore replaced the bare bone in the tube. When it had remained twenty-one hours in the ftomach, it had loft the marrow, and the internal cavity was enlarged, though the girt was leffened. This arofe from the corrofion of the internal and external furfaces at the fame time. Both furfaces were covered with a yellow fluid, that had at once a bitter and falt tafte; and points of gelatinous matter were difperfed over them. The bone, thus half diffolved, was put again into the tube, and left thirty-two hours longer in the ftomach. If the reader will conceive a cylinder of thin paper, uneven at the ends, and perforated

forated with several holes, he will have an idea of the state of the bone when it was taken out of the tube. It was covered with the same fluid, which must have been the gastric liquor; and the gelatinous points now also were dispersed over the surface of the leaf; this jelly was the osseous matter itself, reduced to this state by the action of the gastric fluid. Lastly, nine hours longer continuance in the stomach, left only a few small chips. This one experiment convinced me, that the gastric fluid of the little owl is capable of digesting bone as well as flesh, without the concurrence of any external agent: it also shews the gradual progress of digestion.

CXLIX. Having so far satisfied my curiosity, it remained to enquire into the nature of this fluid, and its effects out of the animal body. With the small spunges, by means of which I obtained so large a quantity from crows (LXXXI, LXXXII), I procured it in due proportion from the species of owl in question. I say in *due proportion*; for it is evident, that as the stomach of these birds will not admit so many tubes as that of crows, it cannot yield so much gastric fluid. Besides, I had only six little owls, whereas I could get as many crows as I pleased. It was wonderful how soon the spunges were filled

filled with liquor. As they were introduced into the empty ftomach, they were foon thrown up, agreeably to an obfervation made above (CXLVII); and yet they were as full of juice, as if they had been dipped in water; and frefh ones immediately forced down the throat, yielded a nearly equal quantity. I obferved the fame thing in crows (LXXXIII). Whence it appears, with what care Nature provides a large fupply of gaftric liquor in thefe animals, as digeftion is entirely dependent upon it. The juice was inftantly fqueezed out of the fpunges into a fmall glafs; it appeared to have the fluidity of water, but was of a reddifh-yellow colour, like the yolk of an egg. This colour was not inherent in the gaftric liquor itfelf; it arofe from an immenfe number of very fmall yellow corpufcles, fcarce perceptible by the naked eye, but eafily feen by help of the microfcope. In a few hours they fubfided to the bottom, in the form of a yellow fediment, and left the fluid above tranfparent, like water, where it has been freed from mud that was diffufed through it and rendered it turbid. The firft time I faw this phænomenon, I fufpected it was owing to fome impurities that remained in the ftomach, and were mixed with the juice. Before the next experiment, in order

to

to be certain that the ſtomach was free from heterogeneous ſubſtances, I kept the animal faſting for a longer time than uſual; but this did not prevent the yellow colour from appearing. Upon opening the ſtomach of an owl that had been long kept faſting, I could find no foreign ſubſtance, but the fluid was as yellow as that ſqueezed out of the ſpunges. I was therefore convinced that theſe particles, though I could not diſcover their origin, did not come from any remains of the food. The gaſtric liquor of the little crow, like other gaſtric liquors, is a little ſalt and bitter: it evaporated ſooner than water. It leaves a ſediment of the yellow particles, which gradually becomes dry, and forms a blueiſh yellow cruſt; it is not at all inflammable. It has one property common to every gaſtric fluid I have hitherto examined, or ſhall have occaſion to mention in the ſequel; though it is expoſed to the open air for weeks and months, in the hotteſt ſeaſon, it never becomes putrid.

CL. Such are the properties of the gaſtric liquor of the ſmall owl, when examined alone. Let us proceed to the effects it produces on fleſh out of the body. In theſe experiments I uſed calves inteſtines, a kind of food which this bird devours very greedily.

dily. Forty-fix grains were immerfed in fome recent gaftric fluid; and at the fame time an equal quantity of the fame inteftine was put into a phial exactly like the former, and an equal quantity of water was poured upon it. Whenever I have made experiments, with a view to compare the effects of the gaftric liquor and water, I have taken care that all circumftances fhould be alike. To prevent evaporation, the mouths of the phials were ftopped with paper; they were fet near a kitchen fire, where the ufual heat was between thirty and thirty-five degrees. In eleven hours fome black fpots began to appear upon the inteftine in the gaftric fluid, which were at firft thinly fcattered over it, but became gradually more numerous, till in twenty-four hours they almoft covered it. During the formation of the fpots, I examined the inteftines with the microfcope, and found that where they appeared, the flefh was foftened, and had loft its fibrous texture. When they had fpread over the whole piece, I took it out of the liquor, and wafhed it with pure water; and now it recovered its white colour, for the black covering confifted of a thin ftratum of flefh, which the gaftric fluid had concocted. It was very eafily rubbed off, and fell to the bottom of the water in exceedingly

ceedingly small particles, where it formed a black sediment, and when viewed by the microscope, seemed to be a collection of molecules of flesh, with no appearance of fibres. When the piece of gut was dried, it weighed only twenty-eight grains, and had therefore lost eighteen; the piece that had stood in water for the same length of time, was quite fœtid; whereas the other emitted no disagreeable smell: after washing and drying, it was found to have lost seven grains. Both pieces were again put into the phials, with the same quantity of water and gastric fluid, and left in their former situation for two days. The latter had now lost its shape and organization, and was converted into a black mucilage, of which the particles had no longer any cohesion. The gastric liquor had therefore dissolved the piece of intestine completely; an effect, which neither water nor putrefaction had produced upon the other; for there was a remainder of nineteen grains, that not only retained its fibrous structure, but made considerable resistance when I attempted to tear it.

CLI. I did not neglect to examine the stomach and œsophagus of this species of owl, as I conceived that it would be improper to omit a brief description of these organs

gans in the animals upon which my experiments were made. If the beginning of the duodenum be tied, so as to stop the air from passing, and the upper end of the œsophagus be inflated, we get a view of the œsophagus and stomach dilated to their utmost extent; together they resemble a pear, or rather a gourd, of which the belly is formed by the stomach, and the neck by the œsophagus; when viewed against the light, the latter appears semi-transparent, and the former quite opake. If they are cut longitudinally, and spread upon a table, we find that the transparency of the œsophagus is owing to the thinness of its sides, which thicken as they descend, and render the lower part as opake as the stomach. It becomes not gradually, but suddenly thicker, from the multitude of follicular glands that form the same kind of transverse fascia, that I have described in other birds; in this species it is about five lines broad. These glands continually secrete into the cavity of the œsophagus a liquor almost insipid, of a turbid white colour, and of some density; in a word, resembling the œsophageal juice of other birds. At the beginning of the stomach the follicles disappear, nor could I find the smallest vestige of any thing like them in the coats, though I searched

searched with care. Are we then to suppose that the fluid, which is always to be found in the stomach, derives its origin from the numberless glands lying at the bottom of the œsophagus? this is probably true of some part of it; but that no small part comes also from the arteries of the stomach itself, the moisture, like what I have described in other animals (XCIII, CXXXII), has furnished me with an indubitable proof; for it immediately appears again, though it has been wiped off ever so clean.

CLII. This description will apply to the œsophagus and stomach of screech-owls: I have made experiments on two species of owl; one variegated with many colours, among which the red and brown, or dull yellow, predominate; upon the head are two curious tufts, in the shape of a crescent; the other species has not this tuft, but is adorned with a greater variety of elegant colours; the iris is dusky, in the former it is yellow (*a*). My first experiment was made upon one of the long-eared owls, and the result greatly surprized me. It threw up two tubes in about three hours after it had taken them, nor was

(*a*) The former species is called by Linnæus *strix otus*, and *moyen duc* by Buffon; the latter *strix studula* and *chat huant*.

the

the flesh at all changed; I could not perceive any alteration, even when it had continued upwards of seven hours in the stomach. If I had not been very cautious in forming opinions, I should have concluded, that the gastric juices of this species are insufficient of themselves to produce digestion; but I reflected, that a single experiment did not warrant such a conclusion, and that some adventitious circumstance might have affected the result. The bird seemed quite stupid, and reduced very much in its flesh; hence it was probably unhealthy, and consequently incapable of digesting its food properly. This suspicion was confirmed by the account of the person from whom I had it, who informed me, that it had refused food ever since it was taken, which was now four days. It was an old bird; and, upon turning to Buffon, I found that, in order to rear individuals of this species, it is necessary to catch them young, for the old ones will not take sustenance in confinement (*a*). In two days and a half longer, that in my possession died without taking any food of itself; it had always returned what I forced down the throat.

CLIII. This owl fell into my hands in winter; the spring following I procured two

(*a*) A. l. c.

young

young ones from the neſt, which devoured food with eagerneſs whenever they were hungry: I now repeated my experiment, and the reſult was exactly the reverſe of the preceding; the fleſh in the tubes ſhewed ſigns of ſolution in three hours and three quarters, and in ſeven was entirely diſſolved. This convinced me, that the failure of the foregoing experiment was not owing to the inefficacy of the gaſtric fluid, but to the morbid condition of the animal; which either leſſened its quantity, or, what is more probable, impaired its quality. I might, therefore, have omitted mentioning that failure; but it was better to relate it, in order to ſhew, that when the food incloſed in tubes is not digeſted, we are not immediately to infer, that the gaſtric fluid is not capable of producing this effect.

CLIV. But my young owls digeſted not only fleſh, but bone; and that of a hard texture, ſuch as the bones of ſheep and oxen, not to mention thoſe of pigeons and fowls. The reſult was eſſentially the ſame as in the preceding ſpecies (CXLVII, CXLVIII); inſtead, therefore, of dwelling upon it, I will relate, at ſome length, a fact, which, in my opinion, deſerves to be noticed. I gave one of my owls a frog, and an hour afterwards killed it. The ſtomach though exceedingly dilated,

dilated, was incapable of containing the whole frog, of which the head lay in the œsophagus, and ſtretched its ſides conſiderably; the hind legs lay at the bottom of the ſtomach, and the fleſh was ſo much waſted, that the bones were nearly bare: the integuments of the thighs and trunk were almoſt corroded, and the fleſh was as tender as if it had been boiled. The head, which lay contiguous to the faſcia of follicular glands at the bottom of the œſophagus, had begun to be diſſolved. This experiment ſhews not only that fleſh is digeſted with great quickneſs by the gaſtric liquor, but likewiſe that it is digeſted equally ſoon in the œſophagus and ſtomach; an obſervation I had not yet made upon any other animal.

CLV. Before I killed both theſe owls, I was deſirous of having ſome of their gaſtric juice, that I might ſee whether it retained like others its power of digeſtion; and I found, that it completely diſſolves fleſh, when it is aſſiſted by a proper degree of heat.

CLVI. In the other ſpecies, the *tawny owl*, the ſame phænomena occurred with reſpect to the ſolution of fleſh and bone in the tubes, whether we conſider the digeſtion of fleſh and bone in tubes, or the ſpeedy digeſtion in the

the œsophagus (*a*), or the remarkable flow-
ness of that process out of the body. Upon
an individual of this species I made an experiment, which had been unsuccessful on the
little owl. Observing, that when they were
hungry and open their beak very wide, if I
dropped a pea, French-bean, or cherry into
it they swallowed it with as much avidity as
if it had been the pleasantest kind of food, I
was desirous of seeing whether the stomach
would digest vegetable substances. * With
this view I enclosed some of the seeds just
enumerated in tubes, and forced them down
the throat, but to no purpose; for though
the liquor swelled them, and perhaps altered
the colour, they underwent no diminution of
bulk. They were thrown up undigested in a
day or two, a circumstance which sufficiently
shews, that such kind of food, notwithstand-

(*a*) When I was writing this passage I was struck by a reflection; for which this is the proper place. If we compare
the present with the LXXVII, LXXVIII, LXXIX, XCIX, C, CI,
CXXXV, CLIVth, it will appear, that the œsophageal before its
mixture with the gastric fluid, in many animals, is endowed
with some degree of digestive power. Though it generally
exerts this power only when mixed with the gastric fluid, yet
in some animals, which swallow their food with great eagerness and have not room enough in the stomach to contain it all,
in consequence of which part must be lodged in the œsophagus,
digestion takes place there.

ing the birds appear to relish it, is ill adapted to their gastric juices. The greediness with which they swallow such substances can arise only from that blind appetite, in consequence of which young birds take whatever is offered them.

CLVII. Being satisfied with these experiments on nocturnal birds of prey, I turned my attention to some of the diurnal ones. My first subject was a falcon given me by my illustrious friend the abbe Corti, formerly professor of natural history at Reggio, and now superior of the College of Nobles at Modena, a philosopher well known in the republic of letters by several fine publications. It was of the size of a common hen, and appeared to belong to the species denominated *lanarius* by Linnæus. I soon found, that I could not handle this bird so familiarly as those which I have had occasion to mention hitherto. Its strong beak and long sharp talons would not easily permit me to open the mouth by force, and thrust the tubes down the throat. I however contrived a method of introducing them into the stomach unperceived by the bird; it consisted in cutting some flesh in pieces, making holes in them, and concealing the tubes in these holes. When the falcon was hungry he ran eagerly to the pieces of
flesh

flefh, and fwallowed them whole. For the fraud to fucceed, it was neceffary that the tubes fhould be quite covered with flefh; for if any part of them was bare, the falcon would put them under his talons and tear the flefh away with his beak and fwallow it without the tubes.

CLVIII. My firft experiment was made with a view to afcertain, whether it was capable of digefting bone independently of the action of the ftomach, which proved to be the cafe; but I have before faid fo much on the fubject of the digeftion of bone, that I fhould forbear to relate the prefent inftance particularly but for a new and important phænomenon, which renders the detail neceffary. The bone confifted of little fplinters of an ox's thigh bone; they were very hard and compact, and of various fizes, from that of a grain of wheat to that of a bean; they weighed together fixty-feven grains; I put them into two tubes, in which they were rather clofely crammed. To prevent their falling out of the tubes when they began to be diffolved, and confequently to get loofe from each other, I put the tubes in a linen bag, a precaution which I had before employed, and continued to employ occafionally in future. In twenty-four hours the bones had fhifted their re-

fpective

spective places and rattled in the tubes, a circumstance that shewed the bulk to be diminished. They were moist with gastric liquor, but had none of those gelatinous points which I had seen in an experiment both on the *little* owl (CXLVIII), and the two other species. These points were, as I then remarked, the osseous matter converted into jelly or chyme by the gastric liquor. But what is extraordinary was, that these splinters retained their original hardness and rigidity; so that at first sight one would not have supposed, that the fluid of the stomach had had any effect upon them. However, the contrary was certain; for when the gastric liquor was wiped off, they weighed only forty-two grains. I now replaced them in the tubes, and examined them again after they had been two days in the stomach. The pieces of the size of grains of wheat were all destroyed but two, which were now no larger than millet. Three of the splinters were at first as big as beans, but now reduced to the size of maize. Those of an intermediate size were diminished in proportion. During the whole time they all continued hard. At the third examination, after fifty-seven hours longer continuance in the stomach, the three large pieces only were left, and they were

now

now not larger than millet; when I ſtruck them with an hammer, I found that they retained their original hardneſs.

The gaſtric liquor therefore of the falcon does not, like that of owls and many other animals, inſinuate itſelf into the ſubſtance of the bone, but acts on the ſurface only. The phænomenon, I think, may be thus explained: conceive a bone to be compoſed, like wood, or to bring a more familiar inſtance, like an onion, of a great number of ſtrata. The ſtrata of the onion are of conſiderable thickneſs, but we muſt imagine, that in bone they are exceedingly thin. The gaſtric fluid of owls or other animals will firſt diſſolve the upper ſtratum, but while it is doing this it will penetrate and ſoften the contiguous ſtrata, without diſſolving them. Hence the tenderneſs of bone that has lain in the ſtomachs of animals. On the contrary, we muſt ſuppoſe, that the gaſtric liquor of the falcon has no power of penetrating the internal ſtrata, but that its action is limited to the ſurface. According to this ſuppoſition the bone will be digeſted without having the internal parts ſoftened, and thus ſtratum after ſtratum will be taken away, juſt as it would happen if we had a menſtruum capable of diſſolving only the ſuperficial layer

layer of an onion without acting upon the others.

CLIX. Before I concluded pofitively that the gaftric juice of this bird does not foften bone at all, I determined to obferve its effects when it is at liberty to act without any obftacle; for it is poffible, that its efficacy might be impaired by paffing through the cloth. I therefore took a piece of the fame thigh bone from the thickeft part, and worked it into a fphere by the lathe, to prevent the angles injuring the fine coats of the ftomach; it was then given to the falcon. My purpofe was to obferve whether as it was diffolved it was alfo foftened.

It continued five days in the ftomach without becoming in the leaft tenderer. The fhortening of its diameter fhewed that it was leffened in bulk. Meantime the falcon threw up the fphere once or twice a day, according as he was fupplied with food; for, as I have obferved with refpect to other birds of the fame clafs (CXLVIII), he did not vomit indigeftible bodies till he had digefted the other contents of the ftomach. To caufe indigeftible fubftances to remain in that cavity after other bodies are digefted, I gave him frefh food; for experience having taught me to judge, when that period was approaching, I

was

was sure to attain my purpose; since when the crop is full of food, the contents of the stomach cannot be evacuated through the mouth. By this contrivance the falcon was made to retain the globe twenty-two successive days. It is scarce worth while to observe, that it was not softened, since the inability of the gastric fluid to produce this effect has been sufficiently proved before; but the remarkable diminution it underwent deserves to be noticed. The sphere was at first four lines and an half in diameter, and when it had been thirty-five days and seven hours in the stomach it measured only a line and about a third; it preserved its form perfectly; the same may be said of its polish; there was not a furrow, nor an indentation, nor an asperity of any sort upon the surface. This smoothness is, I think, a clear proof, that the stomach of this species has no triturating power, otherwise the globe would have sustained some injury from the friction and impulses of so many tin tubes as were introduced into the stomach during its continuance there.

CLX. Let it not however be imagined, that bones of a texture less compact require so much time to be dissolved; this was very far from being the case. My falcon would eat a whole pigeon at once, for birds of this kind when they

they take any large prey, always fill themselves quite full, and then continue several days without food. My falcon refused the entrails, the tips of the wings, and the beak; the rest he devoured with the utmost greediness. But no bone or flesh was ever vomited, nor did any thing pass out at the vent in the form of bone or flesh; the excrements now, as well as at other times, consisting of a semifluid matter, partly white and partly black. When dry it might be reduced to an impalpable powder by rubbing between the fingers. This animal therefore digested not only the flesh, but the bones of a pigeon, and that in the short space of a day; for at the expiration of this time it would eat a second pigeon.

CLXI. While I was examining the manner in which the falcon digests bone, I was struck with a thought that had never occurred to me during the whole train of the foregoing experiments; it was to enquire whether the gastric liquor besides bone is also capable of digesting some other animal substances, such as the enamel of the teeth, the toughest tendons, and horn. With this view I enclosed two incisors from the lower jaw of a sheep in a tube, which the falcon retained three days and seven hours. Wherever the enamel did
not

not extend they were corroded and wafted, but the other parts were uninjured, and as brilliant as at firft. In four days and an half longer continuance in the ftomach the fang was nearly diflolved, but the enamel was perfectly found. The teeth were kept two days more in the ftomach without the tubes, but no further effect was produced; whence it was neceffary to infer, that the gaftric juice of the falcon is incapable of diflolving the enamel of the teeth; a circumftance which is not very furprifing, fince it differs from every other offeous fubftance.

CLXII. I have elfewhere obferved, that birds of prey, and confequently falcons, vomit the feathers of the birds which they eat (LIX); it is therefore evident, that the gaftric fluid cannot digeft them. The fmell emitted by burning feathers fhews, that they refemble horn in their nature; it was therefore reafonable to fufpect, that corneous fubftances would not be diflolved in the ftomach, a fufpicion which was verified by the event. Some pieces of ox's and fheep's horn were as ufual concealed in flefh, and given to the falcon. In a few days they were thrown up entire and uninjured. I have remarked, that the internal coat of the ftomach in gallinaceous fowls is not tender and yielding,

as

as in many animals, but firm and cartilaginous (XXXV, XLVII, XLIX, L). Having frequently obferved, that when burned it exhales an odour very much like feathers and horn, I fuppofed that it would in like manner elude the action of the gaftric fluid, which really happened not only in the thick coats of turkeys and geefe, but in the thin ones of pigeons, blackbirds, and quails. When I gave my falcon the whole ftomach of any of thefe fowls, the other coats were foon digefted, but the cartilaginous remained entire.

In tendons the refult was different; for my experiment I chofe an ox's *tendo achillis*, one of the toughest tendons that is to be found in animal bodies. It was hung to dry in fummer for feveral weeks, and thus became fo hard, that a keen knife would hardly cut it. However, the gaftric liquor of the falcon diffolved it both when it was enclofed in tubes, and loofe in the ftomach.

CLXIII. Moft fhoes have the upper leather of calf-fkin, and the fole of ox's hide. Both thefe fubftances are very readily digefted by carnivorous animals when frefh: this at leaft is the cafe with the falcon; but the contrary happens when they have been tanned. Another fact has warned me how cautious we ought to be in forming general rules in phyfics.

fics. Who would not have concluded from the laſt experiment, that every other kind of leather is alſo indigeſtible? Yet the reverſe happened in ſheep-ſkin dreſſed, and dyed yellow. Some lifts of it were encloſed in tubes, and completely digeſted in ſeven hours.

CLXIV. As I had found the gaſtric fluid of other carnivorous animals incapable of digeſting vegetable matters, it was more than probable, that the ſame thing would take place in the falcon. I however thought, that it would be proper to aſcertain this point by experiment, if for no other reaſon, yet on account of the recent inſtance of the uncertainty of analogical arguments (CXLIII). At the ſame time I was deſirous of determining whether digeſtion is the effect of the gaſtric liquor ſolely, as it ſeemed more than probable. The falcon could very well take ſix tubes at a time: four were filled with various vegetable ſubſtances, ſuch as crumb of bread, chickpeaſe, ſlices of pears and apples; in the fifth and ſixth were encloſed mutton and beef. Upon theſe ſubſtances the effects of the gaſtric fluid were exactly the reverſe. The fleſh was totally diſſolved in twenty-ſeven hours, but the vegetables had undergone no alteration. Two freſh tubes, containing in the middle a bit of fleſh, and at the ſides maſticated

cated bread and boiled peafe and chick-peafe, decided the queftion ftill more clearly. The vegetables were undiminifhed, but the flefh, which was furrounded by them, was entirely deftroyed. Thus the incapability of the gaftric juice to diffolve vegetables, and its efficacy on flefh, were fully proved.

CLXV. By means of little fpunges I procured this fluid fometimes when the ftomach was empty, and at others when it contained fome remains of the food, in which cafe it was always turbid and full of heterogeneous matters, of a cineritious yellow colour, and had not much fluidity. When the ftomach was empty it was fufficiently clear, without any extraneous fubftance, had an intermediate colour between yellow and white, was very fluid, and had a faltifh and bitter tafte. With this I attempted experiments on digeftion out of the body, like thofe I have already fo often mentioned, The refult was not different. I obtained the folution of various kinds of flefh by renewing the liquor from time to time, and by applying a heat of thirty deg. the common temperature of thefe animals. With thefe precautions I moreover caufed nearly the half of a fplinter of a bone of beef, weighing forty-four grains, to be diffolved.

CLXVI.

CLXVI. Having made thefe experiments, in my opinion the moſt intereſting the ſubject admits, my next buſineſs was to examine the ſtomach and œſophagus. However, three hours before I killed my falcon, I fed him, in order to ſee what effect is produced upon the food in the craw. It was in part in this cavity, and part had deſcended into the ſtomach, where it had begun to be decompoſed. It was immerſed in the gaſtric fluid, and this incipient digeſtion had the ſame appearance as it has out of the body. The fleſh in the craw, even that which was upon the point of paſſing into the ſtomach, was only a little diſcoloured; this circumſtance ſhews that digeſtion is performed only in the latter cavity, and that in the craw the food is only diſpoſed to be diſſolved more readily.

CLXVII. When a ligature is made below the pylorus, and air blown in at the top of the œſophagus, this part of the alimentary canal reſembles a large inteſtine about five inches long; a little more than half way down the œſophagus is dilated and forms the craw, though we ſhall find, that it has this name improperly, if we compare it with the craws of gallinaceous fowls, which lie at the ſide of the œſophagus, or rather without it; whereas in the falcon the craw is a continuation

tion of that cavity. If we invert and again inflate the œfophagus, and then examine it in a ftrong light, or with the microfcope, we can perceive an immenfe number of glands from the beginning to the flefhy fafcia, not excepting the craw. If we blow in frefh air, and obferve it again with the glafs, we fhall fee the glands, which are of an oblong fhape, and project a little above the plane of the œfophagus, each emit a drop of liquid; this liquid is fo vifcid, that one of thefe drops may be drawn out into a filament an inch or more long; it is infipid to the tafte. The greateft part of the œfophagus, is full of thefe glands, and is entirely membranous; it only becomes mufcular at the commencement of the fafcia, which in the falcon, as well as other birds, feems to confift only of numberlefs follicular glands, and is above an inch in breadth. Thefe follicles are cylindrical, and are all connected by a fine membrane; they have one of their extremities implanted in the external, and the other in the nervous coat of the ftomach: through the latter, the excretory ducts open and difcharge the fame kind of whitifh and vifcid matter that has feveral times been defcribed as belonging to birds that have ſmall follicles. Thefe glands and follicles abundantly fupply the ftomach

with

with fluid; and though it is sometimes destitute of glandular bodies, yet a liquor continually poured into the cavity by exhalent arteries, forms an addition to that which comes from the œsophagus, as is evident from the moisture which appears upon the sides when they have been wiped dry several times.

CLXVIII. The eagle on which my experiments have been made, belongs to the species called by Mr. Buffon the *common eagle,* because it is found upon most of the high mountains of Europe; it was known to Aristotle, by whom it is called ΜΕΛΑΙΝΑ'ΕΤΟΣ, or the black eagle Hence it has received the denomination of *Falco Melanpetus* from Linnæus, who refers, with whatever propriety, the eagle and falcon to one family. Though some naturalists reckon two species of the common eagle, the brown and the black, I should incline with Buffon and Aristotle to suppose, that there is only one. The difference in colour may depend on the difference of age; for we often see animals of the same species, but of different ages, differ in colour. At the time I was in possession of my eagle I had an opportunity of seeing five others, four dead and prepared, and one living, in the possession of the counts Castiglioni of Milan, two noblemen equally remarkable

markable for politeneſs of manners and ſkill in natural philoſophy. Theſe animals all differed from each other in colour, ſome being of a black more or leſs deep, and others of a darker or lighter brown; yet they agreed in the eſſential characters of the ſpecies. They were all nearly of the ſame ſize, ſomewhat exceeding that of a turkey-cock, their legs and feet were covered with feathers, the nails were black, the feet yellow, the bill blueiſh, and the baſe was covered with a bright yellow cere: ſuch are the characters which, according to the French naturaliſt, the brown has in common with the black eagle.

CLXIX. The ordinary food of my eagle conſiſted of live cats and dogs, when I could procure them. It eaſily killed dogs much larger than itſelf. When I forced one of theſe animals into the apartment where I kept the eagle, it immediately ruffled the feathers on the head and neck, caſt a dreadful look at the dog, and taking a ſhort flight, immediately alighted on his back. It held the neck firm with one foot, by which the dog was prevented from turning his head to bite; and with the other graſped one of the flanks, at the ſame time driving the talons into the body; and in this attitude it continued, till the dog expired, in the midſt of fruitleſs outcries

outcries and efforts. The beak had been hitherto unemployed, but it was now used for making a small hole in the skin, which was gradually enlarged; from this the bird began to tear away and devour the flesh, and went on till it was satisfied. I must not omit observing, that it never eat any skin, or intestine, or bone, except very small ones, such as the ribs of cats and small dogs. Notwithstanding this ferocity, and violent impetuosity in attacking animals, it never gave any molestation to man. I, who was its feeder, could safely enter the apartment where the bird was kept, without any means of confining its movements, and beheld these assaults without dread or apprehension: nor was the eagle at all hindered from attacking the living prey I offered it, or rendered shy by my presence. As it was not always in my power, or at least in my will, to give it living food (for I had not always dogs and cats at hand; and gallinaceous fowls, which were equally acceptable, were too expensive) I substituted flesh which, though it was not so well relished, was not disagreeable. In general, when it had flesh at will, it only made one meal a day. I found, by weighing what it eat, that thirty ounces of flesh served it one day with one another. This species of eagle

is

is provided with a very large craw, which of courſe is the firſt receptacle of the food; and when it was at liberty to eat its fill, this viſcus was generally diſtended to a larger ſize than that of a turkey-cock full of grain. It gradually contracts in proportion as the fleſh paſſes into the ſtomach, juſt as it happens in gallinaceous fowls.

CLXX. Some of the firſt times I obſerved my eagle eat, I was ſtruck by a phænomenon, which conſtantly recurred whenever it took food. After it had ſwallowed a few mouthfuls, a thin ſtream began to flow from each noſtril, and to run down the upper ſide of the beak; at the end they joined, and formed a large drop, which ſometimes fell on the ground, but generally paſſed into the mouth, and was mixed with the food. This drop was continually renewed by freſh ſupplies from the noſtrils, as long as the animal continued to feed, and after that it ceaſed to appear. This liquor was of a ſky-blue colour, had a ſalt taſte, and was nearly as fluid as water. But why does it flow only while the eagle is feeding? and what is its uſe? It flows at that particular time only, I ſuppoſe, becauſe the receptacle in which it is contained is then only compreſſed; and the preſſure ariſes from the motion of the mouth,

or

or the impulse of the food against the palate, near which this receptacle lies. Of the use of this fluid, I candidly own my total ignorance. I suspect, however, that as it is mixed with the food, it serves, like the saliva, to moisten it, and facilitate digestion.

CLXXI. It is commonly thought, and the opinion has the sanction of the best naturalists, that birds of prey, and especially eagles, never drink. What I have observed is, that when the species mentioned in the present dissertation, were left even for several months without water, they did not seem to suffer the smallest inconvenience from the want of it; but when they were supplied with water, they not only get into the vessel, and sprinkle their feathers like other birds, but repeatedly dip their beak, then raise their head, in the manner of common fowls, and swallow what they have taken up; hence it is evident that they drink. For the eagle it was necessary to set the water in a large vessel, otherwise, by its attempts to drink, the vessel was sure to be overturned.

CLXXII. To collect into one point of view every thing relative to digestion, let us examine another opinion, more immediately connected with our subject. It is said by several

several celebrated naturalists and physiologists (*a*), that the eagle, when unable to procure flesh, will feed upon bread. To ascertain this point, I made various experiments. I first set before the bird both flesh and wheaten bread; and finding that it ran towards the flesh, without even casting a look upon the bread, I set only the latter before it, and this after a day's fast, when it must have been pressed by hunger; I did not however attain the end I had in view, and therefore kept it fasting for another day, but still to no purpose. When the bread was set near it, it would just look at it, and then turn its eyes towards some other object. When I had prolonged the fast to the fourth day, the bird ran towards me, as I opened the door of the apartment, but with no other view than to ask for food; I offered it a piece of bread, but in vain, for, without even touching it, it returned to the place where it stood before my coming in. I might have carried the trial still further, but was afraid lest the animal should sink under it.

CLXXIII. I therefore abandoned this mode of experiment, and thought it would be better to make the eagle swallow some bread;

(*a*) *Buffon* Hist. Nat. des Oiseaux. T. 1. *Haller.* T. 6.

for it would either be always thrown up, and then it would be reasonable to infer, that this was an unsuitable kind of food; or in case it should neither be vomited, nor voided unaltered along with the excrements, and the animal should shew no symptoms of uneasiness, we must conclude that it is digested and assimilated. I concealed the bread in some flesh, as I had done in my experiments upon the falcon (CLVII), and had recourse to the same expedient, whenever I was desirous that my eagle should take tubes or other substances. For though this ferocious bird was exceedingly gentle towards me, who was his feeder, yet it might have been hazardous to irritate it; and that would have been unavoidable, if I had opened the beak, and thrust bread down the throat by force. The first portion of bread which the eagle swallowed concealed by flesh, amounted to half an ounce. Indigestible bodies, such as feathers, used to be thrown up eighteen, twenty, or, at most, twenty-four hours after they were received into the stomach. But the bread was not vomited in that period, or a day longer; nor did the excrements appear to be altered or mixed with bread. I then gave the animal a whole ounce, instead of half an ounce of bread, none of which was vomited or voided unchanged at the vent. The fame thing took place, when the quantity of bread was increafed to fix ounces. My laft experiment upon bread, was to fubftitute the cruft inftead of the crumb; but the refult was juft the fame; and notwithftanding the eagle had fhewn fo little appetite for this kind of food, its health did not appear to fuffer. And I was obliged to conclude, that this fpecies of vegetable is digefted, and converted into real nutriment, as well as animal matters. I could not therefore refufe to accede to the opinion of thofe, who affirm that eagles, when much preffed by hunger, will feed upon bread, though mine would not touch it.

CLXXIV. But in what manner is bread digefted in the ftomach of the eagle? Is it by the gaftric juices alone, or affifted by trituration? Is any fuch action exerted by it? In fhort, what is the immediate caufe of digeftion? Thefe queftions are too clofely connected with the object of my enquiry, to be paffed over unnoticed. To begin then with the firft. Tubes employed in my ufual manner, would determine the mode of digeftion. And in the prefent cafe alfo, I obferved what I had before obferved in fo many other animals, that trituration had no part in this function, and that it was the fole effect of

the

the gastric juices. While the eagle retained the tubes, a space that never used to exceed twenty-four hours (CLXXIII), the bread which they contained was completely dissolved. If they happened not to remain long so in the stomach, the gastric fluid had corroded the bread, and given it a yellowish colour and a bitterish taste. Where the action of that fluid had been chiefly exerted, the bread was changed into a gelatinous paste, which had nothing of its original taste.

CLXXV. But the tubes shewed, that the gastric liquor of the eagle dissolves not only bread but Parmesan cheese. This power, possessed by a bird properly carnivorous, of digesting a substance so different from flesh, induced me to try whether it is capable of producing the same effect on other matters, and particularly vegetables. But with respect to the latter, I did not find that the efficacy of the gastric fluid extended any further than bread; for several seeds of the cerealia, both raw and boiled, did not appear to undergo any alteration in the tubes, or when loose in the stomach. It is somewhat surprising, that this should be the case with wheat, when wheaten bread is so perfectly digested. We see at least, that vegetables must be triturated before they can be digefted by the eagle, as well as by gallinaceous fowls (XLV).

The foregoing experiments, and the concurring obfervations of others (CLXXII), fhew, that fome animals, fuppofed to be ftrictly carnivorous becaufe they live always upon flefh, and are 'provided with the moft formidable weapons for feizing and deftroying their prey, may yet, under certain circumftances, change their difpofition and manners, and become frugivorous. Thus we read of animals naturally herbivorous, as horfes, fheep, oxen, gradually quitting their ufual aliment, and learning to live upon flefh (*a*). I too can produce a recent inftance in a young wood-pigeon, a fpecies of bird which is univerfally known to feed upon any thing rather than flefh. By dint of hunger I brought it, gradually to relifh flefh fo well that it refufed every other kind of fuftenance, even grain, of which it is naturally fo fond. Such changes, whether effected by defign or accident, will not excite the fmalleft degree of furprize in thofe who know, that of the various kinds of food ufed by man and animals, the gelatinous part fupplies the nutriment, and that this exifts alike in vegetables and animals (*b*). The

(*a*) Haller, Phyf. T. 6.   (*b*) Ib. T. 1.

example

example of the eagle among carnivorous, and of the horfe, ox, pigeon among frugivorous animals, do not however warrant us to conclude, that the former can be univerfally converted by art or chance into the latter, and reciprocally; for, on the other hand, Reaumur's kite (CXVI) and my owls and falcon (CLVI, CLXIV, CLXVI) were incapable of digefting vegetable fubftances (*a*); not that thefe fubftances are unfit for affording them nourifhment, but becaufe the gaftric

---

(*a*) Mr. Batigne, in his critical refleftions on the experiments of Reaumur, pretends, that we are not to conclude, becaufe vegetables undergo no change in the ftomach of the kite, that the gaftric liquor has no aftion upon them. He fuppofes, that its inefficacy arofe from the vegetables not having been previoufly mafticated. *Premiere Reflexion fur les Experiences de M. de Reaumur.*

But in this Mr. Batigne is miftaken. After I had compleated my differtations on digeftion, I procured a kite of the fame fpecies as that of Mr. Reaumur, and had it therefore in my power to repeat and vary his experiments. I conftantly found, that bread, grain, &c. were thrown up unaltered, both when enclofed in tubes and loofe in the ftomach, though they had been previoufly well mafticated. This fact agrees with my obfervation on the falcon, of which the gaftric liquor could not digeft mafticated crumb of bread. I will add, that an owl, fed with chewed bread alone, died upon the fourth day; and upon diffection, the bread was found in its ftomach undigefted. It is therefore evident, that the incapability of the gaftric liquor of fome animals to digeft vegetables does not arife from the want of previous trituration or maftication, and that this fluid is effentially unfit for diffolving fuch fubftances.

liquor

liquor is incapable of decompofing them, and extracting the nutritious jelly.

CLXXVI. With refpect to the fecond queftion, whether the ftomach of the eagle triturates its contents? I think I have abundant proof, that it poffeffes no fuch power. Not to mention the numerous tin tubes that remained fo long in it without receiving the flighteft injury, I can fafely affirm, that I could never perceive the fmalleft contufion upon the grain (which I gave the bird naked in order to try whether it could digeft it) (CLXXV), whether raw or boiled; in which cafe, the fmalleft compreffion or impulfe would have left evident marks upon the furface. Thefe facts are confirmed by the following obfervation: I took fome ftrips, about a line in breadth and three inches in length, of exceedingly thin fheet lead, and rolling them up in the form of a fpiral, introduced them along with fome pieces of flefh into the ftomach of the eagle, in which they continued eighteen hours. The leaft force would have fufficed to deftroy the fhape of thefe ftrips, and being totally inelaftic, they would preferve whatever alteration or diftortion they might receive from preffure or percuffion. However, when thrown up they retained

their

their spiral form; a clear proof that they had not been subjected to violence of any kind.

Let it not however be supposed, that I mean to exclude motion entirely from the stomach of the eagle. Having frequently found foreign substances within the tubes, and fixed in the perforations, I could not but suppose, that they had been driven into them by some force, and this force could be no other than the agitation of the stomach, which was either extrinsical, and produced by the adjacent viscera, or the periftaltic movement by which the food is expelled through the pylorus. I only assert, that the stomach of the eagle has no action capable of breaking and triturating the aliment, as I think I have abundantly proved. It is likewife clearly ascertained, that the gastric fluid is the efficient cause of digestion by the experiments made with bread and cheese enclosed in tubes (CLXXV); but this will be more satisfactorily shewn by the experiments relative to the digestion of animal substances, which I am now to relate.

CLXXVII. The first thing I wished to know was what changes flesh undergoes in the craw, and I had therefore to contrive a method of getting it back at pleasure. Had this bird been of the same gentle and peaceful disposition

position as gallinaceous fowls, this would easily have been effected; for I should have had only to press the portion of flesh that lay highest in the craw upwards with my thumb and fore-finger, and by a continuance of this manœuvre should have brought it out at the mouth. By this simple contrivance I have often examined grain from the craw of fowls, pigeons, and such birds; but the strength and ferociousness of the eagle altered the case totally. After much reflection, I thought of an artifice essentially the same as that adopted for gallinaceous birds. I gave my eagle only three or four pieces of flesh, of which the last was tied in the shape of a cross with a fine packthread three or four feet long. The eagle, pressed by hunger, devoured the flesh greedily without regarding the string, of which the greater part hung out of the mouth; nor did the bird make any efforts to swallow or throw it up. When I thought it time to examine the piece of flesh I pulled the string forcibly, and the eagle, without growing enraged, opened its beak and allowed me more room for recovering the string, and by consequence the flesh that was fastened to it. Sometimes I used considerable force, but did not succeed, probably on account of the flesh being got too low down in the craw; in this
case,

ease, to free the eagle from the inconvenience, I cut the string close to the beak, and gave it some flesh, which carried down the packthread before it into the stomach, whence it was thrown up in a short time; but I have more frequently succeeded in drawing up the flesh, and thus obtained an opportunity of examining it at leisure. I never could find, that the craw or its juices were capable of digestion. Its weight was nearly the same after it was drawn up as before it was swallowed, nor did it seem as if it was upon the point of being digested; the surface was only a little tenderer, and had lost its redness; it was penetrated with a fluid that was neither salt nor bitter, but quite insipid. Flesh therefore is not digested, it is only macerated in the craw of the eagle, as grain and grass in the craw of gallinaceous fowls.

CLXXVIII. We must therefore conclude, that the whole process of digestion begins and ends in the stomach. If then it was of consequence to know what happens to flesh in the craw, it is of much greater importance to observe how it is altered in the stomach. But as the expedient to get the flesh back, mentioned in the last paragraph, would not be of any service here, I contrived another, which answered wonderfully well. I enclosed

enclosed the flesh in little nets with small meshes, which were generally vomited empty; but in some there were considerable remains of flesh. The pieces I used for these experiments were globular, and the remains almost always retained that figure. They were thoroughly impregnated with gastric liquor, and had both a bitter and salt taste. The surface was gelatinous; when this was removed, the fibres were easily distinguishable, but were as tender as if they had been boiled, and the colour was changed to a reddish blue. When this stratum of tender fibres was taken off with a sharp knife, that below was firmer and less discoloured, and at the center the flesh did not appear to have undergone any change either in its consistence or colour. It is needless to observe that these experiments prove, that the gastric fluid dissolves flesh. The permanency of the globular form clearly shews, that trituration does not take place, but the whole effect, to repeat it once more, is produced by the gastric liquor, which acts upon the surface, and dissolves one stratum after another till the whole is consumed, as we have seen the same liquor of other animals act as well upon flesh as other substances (LXV, CI).

CLXXIX.

CLXXIX. This laſt experiment rendered it ſuperfluous to try, whether the gaſtric fluid of the eagle will diſſolve fleſh encloſed in tubes. Taking this for granted, I proceeded to enquire, whether digeſtion would be retarded in proportion to the toughneſs of the fleſh with which they were filled. With this view ſome of the liver, of the muſcular fleſh of the thigh and heart, a bit of the brain, and a piece of tendon were encloſed in ſo many diſtinct tubes. They continued thirteen hours in the ſtomach, and the gaſtric fluid acted upon them juſt as I had imagined it would. The tube containing the piece of brain was quite empty; of the liver only a very ſmall part remained; the reſiduum of the muſcular fleſh of the thigh was more conſiderable; that of the heart was ſtill greater; but of the tendon there remained moſt of all. Theſe remains of fleſh and tendon had the ſame appearances as I had obſerved in the balls of fleſh that were introduced into the ſtomach without tubes. The gelatinous matter on the ſurface, the tenderneſs of the fibres lying immediately below, and the conſiſtence of thoſe at the center clearly ſhewed, that the gaſtric juices had acted upon the fleſh encloſed in tubes juſt

as

as upon what was left loofe in the ftomach (CLXXVIII).

CLXXX. My next wifh was to know whether its activity would be impaired or deftroyed by paffing through linen before it got to the flefh. With this view, two pieces of the fame tendon and heart, equal in fize to thofe employed in the foregoing experiment, were put into two linen bags, and given to the eagle; in eighteen hours they were thrown up. At firft the fides of the bags were diftended by their contents, but now that which contained the flefh was a good deal collapfed; for half of it was diffolved: the other had more of its original diftenfion; for not above one-third of the tendon was confumed. Upon comparing together the diminution of the fubftances in the bags and in the tubes (CLXXIX), I found, that in the former cafe it was lefs, notwithftanding the bags continued eighteen hours, and the tubes only thirteen. It is therefore evident, that the linen is a greater obftacle to the action of the gaftric liquor than the tubes.

CLXXXI. From my experiments upon crows (LXVII) it was obvious to conjecture, that as more folds of linen were wrapped round the animal fubftances, the action of the gaftric liquor would be ftill lefs confiderable.

derable. I therefore gave the eagle fix bags, containing each an equal portion of beef; the firſt was ſingle, the ſecond double, and ſo on. The bird retained them twenty-three hours, when they were all vomited at once, as uſually happened to tubes and other indigeſtible matters, which when ſmall are thrown up all at once, and when large one immediately after another. The two firſt bags were empty, and the remainder of fleſh in the four others were larger as the folds were more numerous, ſo that in the ſixth it was the largeſt of all. It had however undergone ſome diminution, and the gaſtric fluid had therefore begun to diſſolve it, notwithſtanding the ſix folds, as appeared from its being impregnated with it, and from the tenderneſs of the fibres, and the change of colour on the ſurface. My next wiſh was to try whether the juices of the ſtomach were capable of penetrating through a denſer ſubſtance; I therefore ſubſtituted cloth in the ſtead of linen, and having put ſixty-eight grains of beef in the bag, tied ſome packthread very tight round its neck. In fourteen hours it was vomited, and being apparently of the ſame ſize as at firſt, it was returned immediately into the ſtomach, where it continued twenty-two hours longer. I now

now found that the cloth, notwithstanding its close texture and great thickness which amounted to four-fifths of a line, was thoroughly penetrated by the gastric liquor. The flesh was also moist with it, and appeared, upon being weighed, to have lost twenty-seven grains. Twenty-seven grains had then been dissolved, and as no vestige of them was to be seen in the inside of the bag, it was evident that they must have passed out through the pores of the cloth, and consequently that the gastric fluid is capable of reducing flesh to particles of the utmost tenuity.

CLXXXII. I have before observed, that the eagle devours the smaller bones of dogs and cats along with the flesh (CLIX). When I gave that in my possession a bird, it would also swallow all the bones, except those of the extremities; and as they were not thrown up, there was good reason for believing that they were digested, a circumstance that exactly agrees with my observations on falcons and various other birds (XCVIII, CXLVII, CLIV, CLVIII). But greater certainty was desireable, and this I endeavoured to attain in the following manner: two pieces of the rib of a small dog, each about two inches long, were tied together, and two thigh bones of

a cock;

a cock; this packet was retained twenty-three hours; but the bones were very much altered during that time. The two pieces of rib were reduced to the thinness of a membrane; the least violence was sufficient to break them; they were totally inelastic, and had lost all their marrow. The two thigh bones now resembled tubes of parchment; they were easily compressible, and when left to themselves recovered their shape, and after being bent they would become strait again. Upon one of the tibiæ thus wasted and altered there was a very singular appearance; about one-fifth was still osseous, but tender, yielding to the touch, and much attenuated. It is therefore apparent, that the juices of the stomach are capable of dissolving bone, and that in a short space. I was unwilling to throw aside these bones thus reduced almost to nothing, and therefore tying them up in a bundle I gave them again to the eagle, in order to see whether they would be entirely dissolved, or, like a caput mortuum, retain their membranous appearance; but being apprehensive that this could not be so well ascertained if they were naked in the stomach, I enclosed them in a tube. It was retained thirteen hours, and upon examination was entirely empty; it was therefore reasonable

reasonable to infer, that the gastric fluid had now completed the solution.

CLXXXIII. The readiness with which these bones, of a texture by no means tender, were digested, led me to suppose, that the hardest would not resist the action of the gastric liquor. To determine this, I began by giving the eagle a sphere worked at the lathe out of an ox's thigh bone, of the same diameter as that which had been used for the falcon, and taken from the same individual (CLIX). Upon that occasion I observed, that the falcon did not dissolve it during the long space of thirty-five days and seven hours. In the present case it was every day vomited, and immediately returned, and in twenty-five days and nine hours it was completely digested. The eagle is then capable not only of digesting the hardest bones, but of digesting them in a shorter space than some other birds of prey. In the account of my experiments on the falcon I remarked two things, first, that its diameter decreased without any change of shape; secondly, that the texture was not softened during the whole time (CLIX). The first phænomenon occurred on this occasion, the sphere not only maintained its figure, but continued as smooth as when it came from the lathe. But with respect to the second circumstance, cumftance, there was a wide difference; for notwithftanding the hardnefs of the bone, the furface was fo foft every time it was thrown up, that it was eafy to pare off with a knife, flices as flexible as cartilage. The gaftric fluid then of the eagle, befides diffolving the fuperficial ftrata, had penetrated into the fubftance of the bone and foftened it; an effect which that of the falcon is incapable of producing. Penetrating, however, as it is, it has no action on the enamel of the teeth, any more than that of the falcon (CLXI).

CLXXXIV. We have feen how much more fpeedily the gaftric fluid of the eagle digefts bone than that of the falcon; the fame obfervation may alfo be extended to flefh. The former bird required thirty ounces a day (CLXIX), the latter was fatisfied with twelve, and fometimes with ten. The gaftric liquor of the one then diffolves, in an equal fpace of time, three times as much as that of the other, and confequently the rapidity of digeftion in one is triple of that in the other. I fhould however, upon mature reflection, be inclined to confider this greater rapidity as apparent, rather than real. The eagle indeed digefts three times as much flefh as the falcon in the fame time, but then the gaftric juice of the former is far more copious than

than that of the latter; and if we suppose it to be three times as much, a supposition very admissible, as we shall soon see, every third part will dissolve a quantity of flesh equal to that dissolved by the whole gastric fluid of the falcon. The same remark is applicable to other animals. With how small a quantity of flesh is the little owl satisfied in comparison with the eagle, and consequently how inconsiderable is the solution effected by the gastric liquor; but then how trifling does the quantity of that liquor appear when we consider that of the eagle! The same reflection will recur when we compare a lamb with an ox, or a hare with a horse. But with respect to the case in question, I could not devise any more effectual means of determining whether the greater effect produced by the juices of the eagle arose from the greater abundance solely, or in part also from its superior efficacy, than to give each of these birds a small quantity of flesh at the same time, and observe what would be the event. It would either be digested by one as soon as by the other, and then the same efficacy must be ascribed to both; or else the eagle would digest it more speedily than the falcon, in which case the small quantity of flesh would not allow us to suppose, that the fluid of the

falcon

falcon could not fo foon diffolve it on account of its being in fmaller quantity, and we fhould therefore be obliged to conclude, that it is lefs capable of digefting flefh than that of the eagle. This experiment I have often repeated, not only upon the falcon and eagle, but upon the two fpecies of owls alfo, and crows, and the refult has been, that fometimes one and fometimes another of thefe birds has digefted the fmall portion of flefh fooneft; nor did the eagle at all diftinguifh itfelf above the reft. As the difference of time was very inconfiderable, it may be overlooked, and we may fafely fuppofe, that the digeftive power of the gaftric fluid is nearly equal in thefe feveral fpecies, and confequently that the eagle has no advantage over the reft. It may however be objected, that with refpect to bone, at leaft, the prerogative of digeftion belongs to the eagle in preference to the falcon, which takes above thirty-five days to diffolve the fame fphere which the eagle diffolves in lefs than twenty-fix (CLIX, CLXXXIII). I can adopt this opinion without much reluctance, fince there is no inconfiftency in fuppofing, that two menftrua may agree in the effects they produce upon one body, but differ with refpect to another; nay, this idea is confirmed by the facility

with

with which the gaftric fluid of the eagle penetrates into and foftens bone, a quality which that of the falcon does not poffefs in the fmalleft degree (CLIX, CLXXXIII).

CLXXXV. Let us now proceed to fhew the great abundance of gaftric fluid in the eagle in comparifon with fmaller birds, fuch as the falcon, the owl, &c. To procure this fluid I was not obliged, as in other animals (LXXXI), to ufe fmall fpunges. The eagle fupplied me fpontaneoufly. Very foon after it was in my poffeffion I was aware, that along with the tubes a quantity of gaftric fluid was thrown up, fo that the floor was often quite wet with it. It was eafy to devife a method of catching it before it fell on the ground; for the eagle rarely moved from the place where it took food, and therefore generally vomited the tubes on the fame fpot. Upon this I fet a large glafs veffel, and thus was enabled to collect a large quantity of liquor, which generally exceeded three-fourths of an ounce a day on thofe days when the vomiting took place, a quantity which I could not even hope to procure from all the abovementioned birds of prey taken together. What I obtained in this manner was extremely well fuited to my purpofe, not being adulterated

terated with heterogeneous matters; for it was always thrown up when the ſtomach was empty, as I knew by the avidity which the animal ſhewed for freſh food at this time. Its ſmell, which I cannot deſcribe, is not diſagreeable, but very much reſembles that emitted by the gaſtric liquor of other birds of prey. If we except the colour, which in the others is yellow, but in the eagle cineritious, it exhibited the ſame qualities, whether we conſider the bitter and ſalt taſte, the turbid appearance, which is almoſt inſeparable from the one as well as the other, its fluidity, which comes near to that of water, its diſpoſition to evaporate, or the total want of inflammability.

CLXXXVI. The gaſtric juice of the eagle, as well as that of other animals, is capable of digeſting animal and vegetable matters out of the body. It has even produced an incipient ſolution of bone, and an almoſt complete one of cartilage; but the experiments were made in a conſiderable heat; for otherwiſe little or no ſolution took place, and the gaſtric juices of the eagle now only prevented theſe ſubſtances from becoming putrid.

Upon this fluid I made two experiments, to which I had not ſubjected that of other animals.

animals. On a very cold day in winter I exposed a small quantity in a glass, on a window, along with two other glasses containing water, in one of which was dissolved a quantity of common salt sufficient to give it a stronger taste than the gastric fluid had. The thermometer set beside the glasses stood at 5 deg. below 0 (*a*). Of the three liquors the first that was frozen was the common water, the next was the salt water, and the last was the gastric fluid. When I carried them into my apartment, where the temperature was three and an half deg. above 0, the first that thawed was the gastric fluid, next the brine, and lastly the water. It must therefore be supposed, that this gastric juice is capable of resisting cold more than common water. As this cannot be attributed to its saline principle alone (otherwise it would have been sooner frozen than the brine), it is necessary to admit some other principle capable of retarding congelation, whether spirituous or oily, or of whatever other nature; and the close analogy subsisting between the gastric liquor of the eagle and other animals renders it highly probable, that a like principle exists also in them.

(*a*) Twenty and three-fourths, Fahren.

My second experiment was the following: having learned from Mr. Levret (*a*), that the juices of the stomach have the power of dissolving the inflammatory crust of the blood, I procured some of it from a pleuritic patient, and immersed it in a phial of the gastric fluid of the eagle. The event completely answered my expectation: for in two days and a half, in a temperature of 15 deg. the crust was entirely dissolved, and converted into a liquid of a dark hue: this can occasion no surprize; for if the gastric fluid can dissolve animal substances of a far harder texture, such as muscle, cartilage, bone, out of the body, it will much more easily produce the same effect upon the inflammatory crust of the blood.

CLXXXVII. Here the death of the eagle, which happened somewhat more than five months after it had been in my possession, put a stop to my experiments. I however resolved to examine the parts that are situated internally, the only enquiry relative to digestion that could now be made. During the dissection I found, that this individual was a female; for there were many eggs, some smaller and some bigger, attached to

---

(*a*) Art d'Accoucher.

the

the ovaria. It was confequently much larger and ftronger than the male of the fame fpecies; for it is a conftant obfervation, that the male in birds of prey is about a third fmaller and weaker than the female; whereas, in other cláffes, the male exceeds the female in both thefe refpects (*a*). The inteftinal canal was full of the ufual folds and convolutions; when ftretched out at full length, it was about fifty-nine inches long from the begining of the duodenum to the end of the rectum. There is a double pancreas, and each portion is perfectly diftinct and feparate; but the fame obfervation has been made upon other animals. Both thefe glands are of a blueifh flefh-colour, of an oblong fhape, and fmaller toward the end. There is a difference in the fize, one being an inch and an half in length, whereas the other is only an inch and three lines. They lie parallel, are fituated about five inches from the pylorus, and ftretched along befide the duodenum, one on each fide, and are attached by cellular fubftance. At about fix inches diftance from the pylorus an apparent cord, tinged internally with a dark azure-colour, lies upon the duodenum. If we trace it

(*a*) Buffon. l. c. T. 1.

back-

backwards, we find it gradually enlarged, and at laft inferted in the gall-bladder, which, in fhape and fize, refembles a wood-pigeon's egg. From what has been before obferved (LXXXIV, CXV), it is eafy to guefs the ufe of this cord; it is the duct through which the bile paffes from the bladder into the duodenum. If the gall-bladder be preffed gently the cord becomes immediately tinged with a deeper azure, and the liquor runs into the duodenum: if we open that gut, the upper part is found tinged with a greenifh azure bile. Upon wiping it away, the entrance of the duct becomes vifible, and frefh bile runs into the duodenum when the preffure is renewed. The gall-bladder lies towards the right lobe of the liver, but is not covered by it. The bile is rather denfe, and has a ftrong bitter tafte.

CLXXXVIII. When I infpected the ftomach I was aftonifhed at its fmall fize, when compared with the crop. The latter cavity is capable of containing thirty-eight ounces of water, whereas the ftomach can fcarce hold three. We muft therefore fuppofe, that the great quantity of flefh devoured by this voracious bird paffes flowly from the craw to the ftomach, in proportion as it is

digefted

digested and expelled into the intestines.
Hence it is easy to comprehend how a single
meal may serve several days; for a large prey
will be equivalent to several smaller ones. I
cannot give a better idea of the shape of the
stomach, than by comparing it to a man's leg
and foot. At the point of the toes lies the
pylorus, the foot resembles the bottom of
the stomach, and the leg the upper part. The
fleshy fascia full of follicular glands, which
in other birds, whether granivorous or carnivorous, is situated just above the stomach,
in the eagle is contained within its cavity,
and makes up the superior and larger half.
The internal coat of this fascia is so thin and
delicate, that it tears upon being slightly
rubbed with a cloth. We come next to the
nervous coat full of an infinite number of
pores, out of which, when pressure is made,
issues a viscid, cineritious, and insipid liquor.
Upon removing this coat these pores appear
to be the excretory ducts of the follicles, of
which one extremity adheres to this, and the
other to the muscular coat; next the last
mentioned lies the external coat, which appears to be membranous. The glands are
cylindrical, a line and one-fourth long; they
are tied together by a number of membranous

ous filaments. This short description shews the entire resemblance between the fascia of the eagle and other birds. The four coats pass on to the inferior part of the stomach, and extend to the pylorus. The muscular coat seemed to merit a distinct examination. It consists of two strata. That which lies next the nervous coat is formed by fleshy fasciculi, of a lively red colour, running in a longitudinal direction. The other is of a paler red colour, and the fibres intersect those of the other coat at right angles, and of course run transversely. Notwithstanding their nearness they are perfectly separate from each other, like the rings of certain worms, particularly of the earth-worm, which they moreover resemble in their blueish flesh-colour. These two thin strata doubtless cause the various motions of the stomach, of which the effects have appeared in some of the experiments related above. This coat is one-fourth of a line in thickness; upon the fascia it is thinner, and I could only find the transverse stratum; whence it seems probable, that the motion of the stomach chiefly takes place in the lower part, which has no, at least no apparent, glands; but as a thin transparent liquor oozes out on slight pressure, as

in

in the ftomach of birds belonging to the fame clafs (XCIII, CLI, CLXVII), we muft conclude, that it abounds in fmall arteries, which perform the office of glands.

CLXXXIX. The death of my eagle happened a few hours after it had taken food, but I could not difcover the caufe. Moft of the flefh was in the craw, and a little only had defcended into the ftomach. It lay at the bottom near the pylorus, but fhewed no appearance of being digefted, whether on account of the morbid condition of the animal, or becaufe it had but juft fallen into the ftomach. It was foftened by the gaftric juices, that tafted very bitter, which, as well as its yellow hue, was owing to the regurgitation of the bile into the ftomach, and thefe qualities were more apparent in the vicinity of the pylorus. The flefh in the crop was not altered in confiftence or colour, except that which lay in contact with the fides; this was a little difcoloured and fomewhat tenderer than at firft, circumftances that agree with what was faid at the clofe of the CLXXVIIth paragraph.

Upon emptying, inferting, and then inflating the craw, the furface was covered with a multitude of fmall drops, which, when united

by

by some flat body passing over them, formed a fluid as transparent and thin as water; it seemed to have a bitterish favour. Upon inspecting the places whence the drops arose, they seemed so many points, which the microscope shewed to be minute pores. Hence it appeared, that every part of the crop abounds with these perforations, which I had no hesitation in supposing to be the excretory ducts of a multitude of glands lying between the coats, as I had also found in the craws of other birds (XLIX, L, CLXVII). In search of them I dissected away the internal coat, which, in thickness and strength, resembles the nervous coat of the stomach, of which perhaps it is a continuation. But neither in the substance, or between it and the muscular coat, did I find any appearance like glands. All that I could perceive, when I held the internal coat against the light, was the pores already mentioned, that looked like lucid points. Nor did the muscular or the external coats, which last is membranous, contain any glandular body. I was therefore obliged to conclude, that the fluid oozing out in the form of numberless drops upon the internal surface of the craw is secreted, not by glands, but by arteries too small to be visible.

visible. The rest of the œsophagus is also full of these pores, and the same fluid issues out from them; no small part of which must run into the cavity of the stomach, and contribute to the formation of the gastric menstruum, which is composed of this and the proper fluid of the stomach, of the bile, and the pancreatic juice.

DISSERTATION V.

THE SUBJECT OF DIGESTION IN ANIMALS WITH MEMBRANOUS STOMACHS CONCLUDED. THE CAT. THE DOG. MAN. WHETHER DIGESTION TAKES PLACE AFTER DEATH.

CXC. THE great difficulty with which cats are made to swallow tubes, and the facility with which they vomit them, hindered me from making experiments upon this irritable animal in the manner I could have wished. Among, however, a vast number of unsuccessful trials I have once or twice succeeded, and thus have been enabled to illustrate one chief object of my enquiries, I mean the efficient cause of digestion. I have used every effort to oblige this animal to swallow bread and flesh, their ordinary food,

enclosed

enclosed in tubes, and in two individuals, one an adult, and the other a young one, have forced them into the stomach. Both were killed, one after having retained tubes filled with flesh nine, and the other with bread five hours. The former were found near the pylorus. The outside was wet with gastric juice, the grating at the ends was entire, as also were the tubes, upon which there did not appear any bruise or other injury. Two of the tubes were empty, the third contained a bit of the size of a lentil-seed macerated in the gastric fluid. The center preserved the colour, consistence, and taste of flesh; the surface was changed into a greyish jelly of a bitterish taste.

The tubes containing bread having remained only five hours in the stomach, were not empty. It had been chewed before it was put into the tubes, by which it was moulded into the shape of cylinders six lines and three-fourths long. These cylinders were not completely dissolved, a portion about four lines long remaining towards the middle of the tube, which was externally gelatinous, but internally retained the characters of bread. This experiment then furnishes an irrefragable proof, that the gastric fluid, as well in the cat as in other animals with membranous and

and intermediate ſtomachs, is the efficient cauſe of digeſtion independently of any triturating power.

CXCI. If the ſtomach be inverted and then inflated, it will be covered with humidity, though care ſhould have been taken to wipe it dry. This humidity will appear repeatedly after the ſtomach has been freed from it, a phænomenon common, as we have ſeen, to various other animals. It is not poſſible to diſcover the pores from which this fluid iſſues by the aid of a microſcope, nor can any glandular bodies be perceived in the coats, or the intervals between them; but when the ſtomach is held againſt the light, and examined with a glaſs of great magnifying power, a number of bright flat meſhes or eyes appear through the coats. I could not however determine the nature of them, notwithſtanding I conſidered the different parts of the ſtomach with ſome attention.

CXCII. My ſucceſs with dogs was much greater than with cats. I could make them take more tubes without being liable to the inconvenience of having them vomited. I could not however force them down the œſophagus, for this operation was attended with the ſame danger as in the falcon and eagle; whenever I attempted it, the animal uſed all its efforts to bite me. But as they would swallow them spontaneously, like those birds, I had only to conceal them in pieces of flesh, and throw them upon the floor of the place where the victim of my curiosity was kept. As I always took care he should be hungry, he generally ran towards the flesh, and swallowed it eagerly without mastication; whereas, the cat would keep it in the mouth, and after chewing it for some time, throw out the tubes generally compressed by the action of the teeth, and swallow the flesh only.

I repeated the experiment that succeeded with the two cats (cxc) upon a dog; the animal took six tubes, four full of various kinds of animal substances, as coagulated blood, lights, muscle, and cartilage; the two others contained chewed crumb of bread. In fifteen hours the dog was killed, and the stomach examined; it contained only four tubes; the other two had not been voided, and I supposed they must have passed on to the intestines, where I found them among the excrementitious matter at the beginning of the rectum. Before I describe the appearances in the tubes, I shall say a few words of the juices with which the stomach abounded. As it contained nothing but the tubes, we may consider the gastric juice as pure. It was

of

of a yellow colour, very bitter, almoſt without ſmell, not ſo fluid as water, and totally deſtitute of inflammability. It evidently confiſts of two ſubſtances, one very thin, the other viſcid and gelatinous, which was depoſited after ſtanding a few hours. If the veſſel was ſet near the fire the clear part evaporated, and left a cruſt which was formed by the gelatinous matter.

The two tubes that had paſſed out of the ſtomach were empty, if we except ſome excrementitious matter that had got in through the meſhes of the grating. Of the other four three were empty, nor could I diſtinguiſh which had contained bread and which fleſh. The tough and compact cartilage alone filled part of the tube in which it had been put, but as far as I could judge by my eye, it was half waſted. It exhibited the ſame appearance as on a former occaſion; it was imbibed with gaſtric fluid, and had acquired the ſame taſte, at leaſt on the ſurface. It was ſo ſoft, that it ſeemed to approach more to the nature of membrane than cartilage.

CXCIII. The reſult of this experiment does not coincide with an obſervation in the *Prælectiones Academicæ* of the illuſtrious Boerhaave, publiſhed with a commentary by Haller.

Haller. The passage is so important, that I must quote it as it stands in the original. " Receptum est in hominum opinione quod ossa animalibus subigantur; cum Helmontianis olim sensit Boerhaavius; ut vero certior esset, curam adhibuit, ut observaret, quid cibis fieret in ventriculis animalium valde cibos coquentium, & experimento cognovit non subigi. Dedit cani devoranda intestina animalium, famelicus erat, affatim deglutiit, subegit minime & per extremum intestinum pendula misere post se traxit. Dedit famelico cani ossa butyro inuncta, reddidit furfura, neque quidquam dissolvit nisi quod in aqua dissolvi potest. Dedit carnes, reddidit fibras carnis exsuccas. Dedit ligamenta, ea post tridium nihil mutata egessit (*a*)."

I reserve till hereafter what I have to say on the famous problem concerning the di-

---

(*a*) It is a general opinion that animals digest bone, and Boerhaave once entertained it in common with the followers of V. Helmont. But in order to be certain, he made the experiment upon a species of animal endowed with strong digestive powers. He gave an hungry dog some intestine, but found that instead of digesting it, that it tormented him by hanging out of his anus. Bones dipped in butter were also given to an hungry dog, but he returned the fragments without dissolving any thing more than water is capable of dissolving. When he tried flesh the fibres were voided with their juices expressed, and ligament was voided in three days unchanged.

gestion

gestion of bone by dogs, and now confine myself to that part of the experiment that relates to the intestine, flesh, and ligament. I must candidly own, that the different results obtained by Boerhaave and myself surprized me. My surprize increased when I considered, that the substances were loose in the stomach of his dog, and of course more liable to be attacked and dissolved by the gastric liquor; whereas my tubes must more or less impede its access. Upon further reflection it occurred to me, that perhaps the dog was affected by some internal malady, which might alter the properties of the gastric fluid, as in the owl mentioned in the fourth dissertation, of which the gastric fluid was rendered unfit for digestion by too long fasting (CLII). This reflection however was not quite satisfactory, and therefore to clear up the matter, I thought it would be better to repeat Boerhaave's experiment, and give a dog some intestine, in order to see what changes it would undergo in the alimentary canal. A middle-sized dog accordingly eat four pieces of the colon and ileum of a sheep, and at the same time took two tubes, containing each a portion of the same intestines. The tubes were voided before the time I had fixed for killing the animal, for both were

found

found among the excrements within about eleven hours after they were fwallowed. When the tubes were wafhed clean, I found that the pieces of inteftine were about half digefted. The folvent having acted upon both furfaces had reduced their thicknefs confiderably, but what was left ftill retained its original ftructure. I then wafhed the excrements, and difcovered feveral pieces of inteftine, wafted indeed, as well as thofe contained in the tubes, yet eafily diftinguifhable.

CXCIV. This experiment does not exactly coincide with that of Boerhaave; it is not however totally repugnant to it, fince the pieces of inteftine were not completely diffolved. My long acquaintance with the circumftances attending digeftion gave rife to a conjecture, which I refolved to fubmit to the teft of experiment. The digeftion of thefe pieces of inteftine, faid I to myfelf, was not complete in the fhort fpace of eleven hours (CXCIII); but may it not be fo in a longer time? Is not the quantity of folution in fome meafure proportional to the quantity of time? Does not this appear from undeniable facts related in the foregoing differtations?

In order to verify my conjecture, I had only to contrive a method to prevent the inteftine

from

from paffing fo foon through the pylorus; and this, I conceived, might be done by enlarging the tubes beyond their ufual fize. I got the laft-mentioned dog to take three fuch tubes filled with as many pieces of the large inteftine of a fheep, as amounted to half an ounce and four penny-weights. The tubes were concealed in pieces of the fame inteftine. The dog, which as ufual, was hungry at the time of the experiment and was ftill kept fafting, voided fome excrement in twenty-one hours, and upon minutely examining it, I thought I had fome foundation for believing, that my conjecture was not fallacious; for though fome membranous and fibrous fragments appeared among it, which could be nothing but pieces of the inteftine in which the tubes were concealed, yet they were much more wafted, and much lefs eafily diftinguifhable than in the former experiment (CXCIII), on account of their longer continuance in the body. As the procefs of digeftion is lefs rapid in the tubes than in the open ftomach, I let the dog live twenty hours longer, when the three tubes had remained forty-one hours in the ftomach. They all three lay clofe together at the inferior orifice of the ftomach, wrapped up in fome bits of rag, which in all likelihood the animal had

fwallowed

swallowed before the experiment, and both the tubes and rags were immerfed in gaftric fluid. I make no mention of this juice, having found it to poffefs the properties defcribed in the cxciid paragraph. The reader is more interefted in knowing what happened to the inteftines contained in the tubes. Nothing could have fucceeded better than this experiment: two of the tubes were empty, and what remained in the third did not amount to eleven grains; and thus I had the fatisfaction of finding, as I had conjectured, that the incomplete digeftion of pieces of inteftine fometimes obferved in dogs, is no proof of the inability of the gaftric fluid to diffolve them; it only fhews, that they have not been long enough fubjected to its action. Hence the reafon of Boerhaave's miftake appears evident. Perceiving fome inteftine which he had given to a dog hanging out behind, he concluded that the animal could not digeft fuch fubftances (cxciii); whereas from the facts juft adduced it is obvious, that they had only not continued in the ftomach a fufficient length of time.

cxcv. Thefe facts alfo fhew, that flefh lofes its fibrous ftructure in the ftomach of the dog, and it retains only when it happens to be voided foon after it has been fwallowed.
But

But it might be objected by a rigid partizan of Boerhaave, that I have not strictly proved, that the solution extends to the fibres; for they may have been gradually separated from the common mass, and passing out at the pores, and especially through the meshes of the lattice-work, have left the cavity of the tube empty. I therefore thought it would be proper to throw further light upon the subject by a decisive experiment. If pieces of flesh were inclosed in purses of very thick linen, they would either be dissolved and leave no vestige behind them, as by other animals in like circumstances (LXVII, CLXXX, CLXXXI); and in this case we must infer, that dogs are capable of dissolving flesh completely, or else the fibres would remain in the purses, and then we should be obliged to agree with Boerhaave, that the digestion of flesh by dogs, consist in expressing the juices and converting them into chyme, while the solid parts remained unaltered. But along with the flesh I subjected harder and more tenacious animal substances, such as tendon and ligament, to the test of experiment. Six bags of very thick linen were given to two dogs; four containing four different sorts of flesh, viz. beef, veal, horseflesh, and mutton; and the two others tendon and ligament of an ox. Each

bag

bag contained a quarter of an ounce; and it is to be obferved, that the contents were not cut into fmall pieces. Being apprehenfive left thefe bags, though of fome bulk, would pafs through the pylorus before the time for examining them, I tied to each a bit of dry fpunge. They continued four days in the ftomachs of the two dogs; but fearing left fo long a faft might be hurtful to the animals, and of courfe difturb the procefs of digeftion, I fed them feveral times, though rather fparingly. At the expiration of the time juft mentioned, I killed, and immediately opened them. The experiment fucceeded juft as I could have wifhed; for all the bags were in the cavity of the ftomach. Sufpecting that they might have been torn by the teeth of the dogs, I took particular care to examine them, but they were whole, and the four firft upon being cut open were as empty as if they had never contained any flefh, but of the tendon and ligament there remained about the fize of a hazle-nut; there was no hole in either of the bags. The tendon appeared, upon being weighed, to have loft three-fourths, and the ligament above one half. I examined with particular attention, whether this diminution of bulk and weight arofe from the expreffion of the juices, but the contrary was evident;

for

for they were as pulpy and moift as at firft. Hence I had every reafon for concluding, that the gaftric fluid had really attacked and diffolved the folid parts, fo as to enable them to pafs through the pores of the linen. The fame thing had happened to the flefh. The folution was ftill further confirmed, by the condition of the external ftrata of the tendon and ligament, which were become fo tender as to be torn by the flighteft violence. I was thus fully convinced of the efficacy of the gaftric fluid of the dog in diffolving the fibres both of flefh, tendon, and ligament, though the procefs is lefs rapid in the latter on account of their greater tenacity and harfhnefs. That Boerhaave's dog fhould void the ligaments unchanged on the third day *(ea poft triduum nihil mutata egeffit*, CXCIII), if by this expreffion he means that they retained the nature of ligament, and it feems incapable of any other interpretation, I have not the fmalleft fcruple in believing, having feen the fame thing in a ligament that had been four days in the ftomach of a dog. It had indeed undergone a confiderable diminution, a circumftance which the celebrated Dutch phyfician would alfo have obferved if, inftead of judging by his eye, he had taken the pre-

caution

caution of weighing it before it was fwallowed, and after it had been voided.

CXCVI. We now come to confider whether dogs are capable of digefting bone; a problem which, if we may rely upon the obfervations of feveral celebrated phyfiologifts and phyficians, would appear to be decided in the negative. We have already feen that Boerhaave's dog, after having eaten bones dipped in butter, produced no other change upon them than fimple water would have done (CXCIII). This he alfo endeavours to confirm by the following remark: " Deinde in ftercore canino, quod *Album Græcum* vocant, fragmenta offium pene non mutata reperiuntur, & fit mera offium rafura, quæ dentibus canis adrofit, exfuccorum & in unam maffam reductorum." It appears from fome notes on this paffage and from his great work (*a*), that his illuftrious difciple, Albert Haller, adopted the fame opinion. Dr. Pozzi alfo afferts in his work quoted above (XIII), that dogs do not digeft bone. Of the two experiments which he relates, the following appears the moft conclufive. He gave a dog that had been kept fafting for five days three bones which, though dry, the animal fwallowed for the fake of the butter with which

(*a*) El. Phyf. T. 6.

they

they were anointed. One of the bones weighed three ounces, another two, and the third one. In three days they were voided, and had only loft fix grains.

Such are the ftrongeft arguments adduced by phyfiologifts againft the vulgar opinion. This opinion however found in Reaumur an able advocate, one who eminently poffeffed the difficult art of making experiments with fuccefs. Of the feveral productions by which he has fignalized himfelf, none have contributed more to his reputation, than the two beautiful memoirs on digeftion, which I have fo often quoted with applaufe. To illuftrate the prefent curious and interefting fubject he made the following experiment (*a*). Two compact cylindrical bones, each feven lines long and two in diameter, were given to a fmall bitch, which was killed in twenty-fix hours. The bones that were ftill in the ftomach were diminifhed in bulk, and it appeared to him that feveral laminæ were removed. They had moreover acquired the flexibility of horn, though they were at firft rigid and inelaftic. Hence he infers, that they had been in part diffolved by the gaftric fluid.

(*a*) Mem: 2.

CXCVII. Having noticed the experiments of others, let me relate my own. In the stomach and intestines of the dog, mentioned in the CXCIId paragraph, I found several pieces of bone. They seemed to belong to some quadruped, probably a sheep, and must have been eaten before the dog came into my possession. I did not weigh them, but as far as I could judge by inspection, they amounted to above six ounces. Upon washing and then examining them attentively, I perceived several scars and longitudinal furrows, of which I doubted whether they were produced by the gastric liquor or the teeth of the dog; besides, many of the angles and edges were evidently blunted, at the sight of which the idea of what happens to the hardest bodies in a muscular stomach recurred. I moreover observed, that these places were not so hard as the thicker parts of the bone. But these phænomena only suggested doubts, which I resolved to dissipate by the light of experiment. I therefore filled some tubes with pieces of bone, and gave them to a dog. The bones were of various kinds and degrees of hardness: the tubes, which were two in number, were put into a bag of linen, in order to prevent the bits of bone from getting out. To allow the gastric juices a proper time for producing

ducing their effects, the dog was suffered to live for seven days, and during this time was fed moderately. One of the tubes, though they were rather large, had passed into the cœcum, and was surrounded by feculent matter; the other remained in the stomach. They were neither of them empty, but their contents, which at first weighed three-fourths of an ounce and eighteen grains, now amounted only to four penny-weights and seven grains. All the angles and edges were destroyed. The softest bones had suffered most. They could now be easily cut in the thin places with a knife. The thinnest parts of the bone were dissolved and had passed through the bag, not a vestige of them remaining. This experiment proves two propositions, viz. that the digestive powers of the dog act upon bone as well as flesh; though the latter, on account of its softness, is more speedily dissolved, and that the gastric juices are the sole efficient cause of digestion.

cxcviii. This experiment was repeated three times, and the result was essentially the same; but there occurred two circumstances that deserve to be noticed. One of the dogs produced only a small diminution of the bones in eight days, though it was fed plentifully, and seemed to be in perfect health the whole time.

time. This shews, that little or no effect being produced upon the bone, which sometimes happens, as in the cases alledged by Boerhaave and Pozzi (CXCVI), is no proof of the inefficacy of the gastric fluid of this animal; it only shews, that the digestive powers are unequal; nor ought this to excite our surprize, as the same thing is observable in our own species.

The other remarkable circumstance, to which I alluded, is the reverse of the preceding. Among the bones given to one of the dogs, were two dentes incisores from the upper jaw of a sheep. It has been already observed, that the enamel of the teeth receives no injury from the gastric fluids that are capable of dissolving the hardest bones, such as those of the eagle and falcon (CLXI, CXXXIII); yet the gastric fluid of the dog in question, damaged this compact substance. I have now before me the two teeth, which I keep as a curiosity. The enamel of one is corroded in two, and of the other in three places; the five cavities are above a line long, and penetrate to the nucleus of the bone. The roots of the teeth were almost entirely destroyed. The powerful menstruum of the stomach, had made greater havock among the other bones that were enclosed along with the

the teeth; the excavations as they were wrought in a tenderer fubftance were more confiderable. Upon comparing this phænomenon with the furrows mentioned in the cxcviith paragraph, I have no doubt but they were occafioned by the gaftric folvent. It deferves to be remarked, that in the cafe where the enamel of the teeth was deftroyed, the linen bag had not fuftained the fmalleft injury, though the folvent neceffarily paffed through it: nor is this to be wondered at; for we have inftances of many gaftric fluids that are capable of decompofing the moft compact animal fubftances, though they do not produce any effect on the fofteft vegetable matters (CXLVI, CLVI). This is alfo true of chemical menftruums; the nitrous acid diffolves the hardeft calcareous ftones, but leaves the moft friable gypfum and clay untouched.

CXCIX. Though my experiments on dogs decifively prove, that digeftion is the effect of the gaftric fluid alone; yet it was proper to enquire, whether the fides of the ftomach have any motion during digeftion, and what that motion is? There were two ways of making this enquiry; mediately, that is, by the effects; and immediately, that is, by opening the abdomen and infpecting the ftomach.

With

With respect to the first mode, though I was certain that the stomach of the dog had no considerable motion, because neither the tubes nor the bags had sustained any injury; yet in order to see whether it has any motion at all, I gave a dog some thin tubes open at the ends, which were therefore liable to be compressed by the smallest violence. But I could not find the least contusion upon them, after they had been three days in the stomach. The inspection of the tubes, however, presented a phænomenon which shewed, that the sides of the stomach had not been inactive all the time. Upon opening this viscus I found a mass of hairs, which were of a different colour from those of the dog, and could not therefore have been swallowed while the animal was licking itself. They must have belonged to some other animal that had been devoured by the dog, before it fell into my hands. Many of these hairs had likewise got into the tubes; which must have been effected by the action of the stomach.

CC. I opened five living dogs, taking care not to wound the stomach. This operation was performed soon after they had taken food; for I presumed that the muscular fibres, irritated by the distension, would contract more evidently at this time. The stomach

stomach of the first was perfectly quiescent when it was left to itself. But when the point of a knife was drawn over it, the parts that were touched and those that were adjacent immediately contracted, and then returned to their former situation. Upon throwing round a ligature above the cardia and below the pylorus, and taking the stomach out of the body, I thought I perceived a flight peristaltic motion, but it was of short duration. The contraction and dilatation continued to succeed each other in the places that were touched with the knife, or any irritating body, for half an hour. The stomach of the second was not only destitute of spontaneous motion, but was insensible to every stimulus. In the third stomach the peristaltic motion was very conspicuous; the contraction began just below the superior orifice, and proceeded with a gentle undulation to the pylorus, and the dilatation regularly followed. This spectacle lasted for seven minutes. And I could resuscitate the motion by irritating the upper part of the stomach, but it continued only a little while. The peristaltic movement did not appear on the stomach of the fourth dog, but irritation would excite it. And it was in this case confined to the ring or circular band corresponding to the place where the ſtimulus was applied. This band contracted gently, and the diameter of the ſtomach was ſenſibly diminiſhed; in a few minutes it dilated juſt as ſlowly. In the fifth ſtomach the periſtaltic motion was as apparent as in the third; it laſted ſome minutes longer, and when it had ceaſed in all the other parts of the viſcus, a band juſt above the pylorus continued theſe alternations. The contraction was ſo conſiderable, that the oppoſite ſides of the ſtomach almoſt touched each other; but all theſe motions were exceedingly ſlow, nor did the ſides of the ſtomach ever dilate or contract ſuddenly or forcibly.

CCI. At the ſame time I examined the ſtomach of ſome cats in the ſame manner. The reſult was exactly alike. A gradual movement of contraction and dilatation, beginning at the upper end and extending to the lower, was generally perceptible.

All theſe experiments, and the reader will find ſimilar ones in Haller, though made with a different view (a), clearly ſhew, that the motion of the ſtomach of the dog and cat are not capable of triturating the food, but calculated to carry it ſlowly from the ſu-

(a) Mem. ſur les part. irrit. & ſenſib. T. 1.

perior to the inferior orifice, and thence expel it into the duodenum.

From the great number of dogs that were subjected to thefe experiments I collected a large quantity of gaftric fluid, and found it as capable of producing an incipient digeftion out of the body, as that of the other animals mentioned above, both of boiled and raw meat, and likewife of feveral vegetables. It was however neceffary to apply a pretty ftrong heat, and to change the liquor feveral times, as in other inftances.

CCII. Blafius, in his laborious and accurate anatomy of the dog, fays, that the internal coat of the ftomach is compofed of a congeries of glands (*a*). My opportunities of afcertaining this have been frequent. I have examined it with my naked eye and with the microfcope, but could never perceive any glandular appearance. Upon wiping it dry and preffing it, it was covered with an aqueous exfudation, but I could not diftinguifh the pores from which this exfudation iffues. I have examined feveral pieces with the folar and the fimple microfcope, and in fome perceived a vaft number of lucid points, while in others there appeared nothing of this kind. I then examined the back part, which is con-

(*a*) Anat. Anim.

tiguous

tiguous to the nervous coat, and immediately faw, that it is compofed of a congeries of oblong particles, of a pale flefh-colour, clofely compacted together. Thefe are probably the glands of Blafius; but I cannot affirm that they are really glands, not having been able to diftinguifh the characteriftic marks of glandular bodies in them. But however this may be, it is certain they are deftined to tranfmit a fluid into the ftomach; for whenever they are preffed, the above-mentioned exfudation appears upon the internal furface. And this fluid may be expreffed feveral days after the ftomach has been taken out of the body.

I have before faid, that the pores from which the gaftric liquor iffues are invifible; but the parts contiguous to the pylorus muft be excepted, in which they are very confpicuous. Upon comparing the fluid that thus oozes out with that which is collected in the ftomach when it is opened, we fhall find a very ftriking difference. The latter, as we have feen above, is yellow, bitter, and fomewhat gelatinous (CXCII). But the former has not one of thefe properties, being colourlefs, infipid, and very fluid. Hence it is evident, that the gaftric liquor of the dog, that liquor which is the efficient caufe of digeftion,

geftion, confifts, as in other animals, of feveral different principles, viz. of faliva, of the œfophageal juice, of that which is peculiar to the ftomach, of the pancreatic juice, and of bile.

CCIII. To complete my refearches on animals with membranous ftomachs, it remained to examine that of Man. One may indeed draw very plaufible inferences concerning human digeftion, from obfervations on the other fpecies of this numerous clafs; efpecially from birds of prey, the cat and dog, which refemble us fo much in the ftructure of the ftomach. But analogical arguments are probable indeed, but not conclufive. And it is an object of much higher importance to attain certainty in Man than in animals. In the writings of antient and modern phyficians no topic is more frequently difcuffed, yet there is little elfe befide fuppofition: direct experiments upon Man are entirely wanting, and their refearches are illuminated only by the twilight of conjecture, and fupported by precarious hypothefis. If therefore it was neceffary on other occafions to have recourfe to experiment, on the prefent it was abfolutely indifpenfible. Upon reflection it appeared, that the principal experiments were reducible to two heads, viz. to procure human

man gaftric fluid, in order to examine it in the manner that of animals has been already examined; and to fwallow tubes full of various vegetable and animal fubftances, in order to fee what changes they undergo in the ftomach. I will candidly own, that the latter kind gave me fome apprehenfion. The hiftories of indigeftible fubftances occafioning troublefome fymptoms, and being vomited after a confiderable time (*a*), occurred to my mind. I alfo recollected inftances where fuch bodies had ftopped in the alimentary canal. Other facts however where the refult was contrary, and of more frequent occurrence, gave me confidence. Thus we every day fee the ftones of cherries, medlars, plums, &c. fwallowed and voided with impunity. This confideration at laft determined me to make a trial with as great caution as poffible.

CCIV. I fwallowed in the morning fafting a linen bag, containing fifty-two grains of mafticated bread. All the following experiments were made under the like circumftances. I retained the purfe twenty-three hours without experiencing the fmalleft inconvenience, and then voided it quite empty.

(*a*) Haller, Phyf. T. 6.

The

The string used for sewing and tying it was entire, nor was there any rent in the bag itself. Hence it is plain, that it had not received any damage either in my stomach or intestines. The fortunate result of this experiment gave me great encouragement to undertake others. I immediately repeated it with two of the same bags, with this variation, that one was double, and the other had three folds. My motive obviously was to see, whether these additional folds would impede digestion. The bags were voided in twenty-seven hours, and the double one was empty; but the other still contained a small quantity that had yet the characters of bread.

ccv. From vegetable I proceeded to animal substances. In a single linen bag sixty grains of boiled pigeon were enclosed, and in another the same quantity of boiled veal; both previously masticated. The purses were voided in eighteen hours and three-quarters, and the flesh was entirely consumed. Instead of sixty I next took eighty grains, of which the bulk was not so great as to make me apprehend any danger from its stopping in the œsophagus, and still less from its not getting out through the pylorus, as at that time it must of necessity be very much diminished in bulk. The flesh had been previously boiled and

and mafticated. I retained it twenty-nine hours, at the expiration of which time there remained eleven grains undiffolved. This flefh differed in appearance from that which is taken undigefted out of the ftomachs of animals. The furface of the latter is gelatinous, but the former was as void of fucculency as if it had been fet under a prefs. This appearance, which is analogous to that of the bread in the preceding experiment (CCIV), made me fufpect, that perhaps the human ftomach might poffefs a power of compreffing its contents, though others of the fame ftructure are deftitute of fuch a power. I therefore determined to bring this fufpicion to the teft of experiment.

CCVI. Finding that I could digeft dreffed meat that had been mafticated, I wifhed to know whether I was capable of digefting it without maftication. I fwallowed eighty grains of the breaft of a capon, enclofed in a bag. The bag was retained thirty-feven hours. So long a fpace had produced confiderable effects, for it had loft fifty-fix grains. The furface of the remainder was dry, but the internal fibres appeared to be more fucculent. Digeftion feemed to have gone on uniformly, for the piece retained its original fhape.

CCVII. I

CCVII. I next wished to know whether this dryness of the surface would be observed in raw as well as dressed flesh. I did not doubt but I should digest it in this state more or less speedily; for the human stomach is adapted to the digestion of the one as well as the other, whole nations living upon raw flesh, and raw fish being eaten in some maritime countries; not to mention that oysters, cockles, &c. in the state they are taken, are among the delicacies of the elegant and luxurious, though a food of difficult digestion. I took fasting fifty-six grains of raw veal and as much beef, enclosed in two bags, which were returned about the middle of the next day. Of the veal, as it was the tenderer, there remained fourteen grains, and of the beef twenty-three. In both there was the same dryness on the surface as if the bags had been wrung, or pressed by some external violence.

CCVIII. As then this phænomenon is constant, are we to suppose, that the digestion of flesh and bread, which is produced by the gastric fluid within the bags, is aided by the triturating power of the stomach? Does any such power exist at all? I could devise no better means of solving these doubts, than by observing what happens to animal and vegetable

table fubftances enclofed in tubes. Should they either not be at all or imperfectly digefted, we muft infer, that there was wanting fome circumftance either neceffary, or at leaft expedient; and we might prefume, that it is trituration. I was then under the neceffity of fwallowing tubes. Having fuffered nothing from the former experiments, I entered upon thefe without much apprehenfion. Inftead of tin I had my tubes made of wood, fearing left the refidence of the metal in the ftomach and bowels fhould be productive of bad confequences, although I never perceived any in other animals. The gaftric fluid had never corroded it, the furface was only turned black. My wooden tubes were five lines in length and three in diameter. The fides were, as ufual, perforated with a great number of holes, in order to allow free ingrefs to the juices of the ftomach, along the whole length of the tubes, as well as at the ends. To prevent the entrance of the fœculent matter during their paffage through the long track of the inteftines, they were enclofed in linen bags, a precaution not always employed upon other occafions of the like nature. At firft I took a fingle tube, containing thirty-fix grains of boiled veal previoufly mafticated. The tubes was voided empty in twenty-two hours.

hours. The cover of linen was entire, and had prevented any extraneous matter from getting in.

CCIX. This experiment, which is by no means favourable to the doctrine of trituration, induced me to attempt others before I drew any conclusion. As the tube was capable of containing above thirty-six grains, I put in forty-five. I retained it seventeen hours. There was a refiduum of twenty-one grains; and now appearances were changed; the veal not only had its natural fucculence, but the furface was foft and gelatinous, the center alone remaining fibrous. The jelly was fweet, its fmell was not at all putrid, any more than that of the refiduums in the purfes. Thefe appearances were obferved in three other experiments with boiled, and one with raw flefh of feveral different kinds. I hefitated not to conclude, that in Man, as well as numberlefs other animals, the gaftric fluid digefts the food without the concurrence of trituration. It is indeed not poffible that it fhould concur; for I have direct proofs, that no mufcular action capable of producing fuch effects is ever exerted by the human ftomach. Among the wooden tubes employed in thefe experiments, I procured fome to be made fo thin that the flighteft preffure would crufh
VOL. I. T them

them to pieces; and though I frequently ufed them, not one was ever broken. If I took off the linen cover, which was always entire, and examined them with ever fo much attention, I could never perceive the fmalleft fiffure.

ccx. Thefe perfectly coincide with the following facts. Cherries and grapes are faid to be voided entire (*a*). I refolved to afcertain by my own experience the truth of thefe obfervations. I firft fwallowed four unripe grapes, becaufe in that ftate they have greater firmnefs. In a day they were all voided with the fkin whole; the colour was changed from a greyifh white to yellow. I next made trial of ripe grapes, which, as every one knows, burft on the flighteft preffure. Of twenty-five which I fwallowed eighteen were voided entire, of the other feven the fkins only appeared. I made the fame experiments with many cherries, as well ripe as unripe, and by far the greater number were voided entire. Thefe experiments, together with thofe made on the thin tubes, afford the moft concluíive evidence, that no triturating force is exerted by the human ftomach.

I fhall be perhaps afked, what is the caufe of the drynefs of the fibres, fo often obferved

(*a*) Haller, Phyf. T. 6.

in flesh enclosed in the linen bags, which would appear to have been forcibly pressed (CCIV, CCV, CCVI, CCVII)?· Upon considering the matter I was led to suppose, that the intestines are more concerned in this phænomenon than the stomach. While the bags remain in the stomach, the flesh is converted into a gelatinous matter; for there is no reason to believe that this happens in the tubes only, and not in the bags. But when they are protruded into the intestines, they must be surrounded and pressed by the fœculent matter. Hence the jelly is squeezed out, and the fibres lose their succulence. And this, not the action of the stomach, I take to be the reason why cherries and grapes are now and then burst.

CCXI. Having thus established this fundamental proposition, viz. that the digestion of flesh and bread is produced in my stomach by the gastric fluid independently of trituration (CCIV, CCV, CCVI, CCVII, CCVIII, CCIX, CCX), I had before me a fine field for experiments that could not fail to suggest some important truth. The necessity of mastication is sufficiently known. There is, perhaps, no person who has not some time or other been subject to indigestion for want of having chewed his food properly. In the courfe of my experiments I had fwallowed fome mafticated flefh, and fome without maftication; but having never taken care that it fhould be of equal fize, I had no term of comparifon, and hence was not certain which was moft fpeedily digefted. I therefore fupplied this omiffion in the following manner. I took two pieces from a pigeon's heart, each weighing forty-five grains, and having chewed one as much as I ufed to chew my food, enclofed them in two tubes, and fwallowed them at the fame time, but without attaining the end I had in view; for the tube containing the chewed flefh was voided in twenty-five hours, and the other in thirty-feven, both empty. Another experiment made under the fame circumftances fucceeded better, both the tubes were voided in nineteen hours, and I then faw how much digeftion is promoted by maftication. Of the mafticated flefh there remained only four grains, whereas of the other there were eighteen left. This was confirmed by two other experiments, one made with mutton, the other with veal. The reafon is obvious. Not to mention the faliva which moiftens the food and predifpofes it to be diffolved, it cannot be doubted, that when it is reduced to pieces by the action of the teeth, the gaftric fluid penetrates more readily,

readily, and by attacking it at more points, diffolves it more fpeedily than when it is whole. This is true of menftrua in general, which always diffolve bodies fooner when they have been previoufly broken in pieces. This is alfo the reafon why in other experiments, mafticated bread and dreffed flefh were more readily diffolved than unchewed bread and raw flefh. The boiling had made it tenderer, and confequently difpofed it to allow ingrefs to the gaftric fluid.

CCXII. It is an opinion common among modern phyfiologifts, that flefhy fibres, tendon, cartilage, and bone lofe their juices in the human ftomach, but that the folid parts are not diffolved or digefted. With refpect to flefhy fibres, I muft differ from them, having clearly proved the contrary by experiment (CCV, CCVIII, CCXI). As I could bring the other fubftances to the fame teft, I would not neglect an enquiry of fo much importance. I at firft took membrane enclofed in a tube without maftication or divifion, weighing about fixty-five grains. The tube was voided in thirty-two hours, and prefented the following appearances: The membrane was entire, but feemed thinner and fhorter. It weighed only twenty-eight grains. This diminution, however, was not

a fufa sufficient proof of the solution of the solid parts; it might depend upon the privation of the fluids. It was therefore proper to return it into the stomach, and wait the result. The membrane was voided in fifteen hours; it was still in one piece, but exceedingly reduced, weighing now only five grains. This petty remainder I swallowed a third time; the tube was voided in twenty-two hours, and was now completely empty. I afterwards observed the same phænomena in membranes of greater thickness and tenacity; and I once digested the aorta of a calf after it had been boiled. The only difference I could perceive was, that the compacter membranes required more time to be dissolved.

CCXIII. I made experiments upon cartilage and tendon at the same time. To avoid disgusting the reader by too particular a recital, I will only mention the bare results. The cartilage was more speedily dissolved than the tendon, the former being totally consumed in eighty-five and the latter in ninety-seven hours. Both were taken from an ox, and had been previously boiled for half an hour.

CCXIV. Bones still remained, and I submitted some both of a hard and soft texture to experiment. The latter were completely dissolved,

diſſolved, and required about the ſame time as cartilage. But the former underwent no perceptible diminution, though it continued upwards of eighty hours in my ſtomach. I likewiſe ſwallowed a naked ball of hard beef bone three lines in diameter, and in thirty-three hours voided it undiminiſhed.

It is therefore certain, that the ſtomach is capable of digeſting not only muſcular fibres, but membrane, tendon, cartilage, and even bone itſelf, provided it is not too hard; though moſt phyſiologiſts and phyſicians have been led to adopt a contrary opinion by obſerving, that theſe ſubſtances are evacuated unaltered. But this is no proof that they are indigeſtible (for if they had made the experiment on themſelves, and weighed the ſubſtances, they would have obſerved a waſte), it only ſhews, that they are not ſo ſoon digeſted as other kinds of food, which are diſſolved in a few hours; whereas, membrane, tendon, cartilage, &c. require ſeveral days, on account of their tenacity and hardneſs.

Let no one ſuppoſe, that my ſtomach, being ſtronger than common, is capable of digeſting what that of others cannot digeſt. I own, with concern, that it is weak, as is uſual in thoſe whoſe purſuits condemn them to a ſedentary and unwholeſome way of life.

My ftomach digefts food fo flowly, that I cannot ftudy for five or fix hours after a fparing dinner, and am liable to indigeftion whenever I feed more plentifully than common.

Before I quit this fubject let me obferve, that though I have mentioned the gaftric juices as the efficient caufe of digeftion in the experiments on myfelf, yet I mean not to exclude thofe of the inteftines from their fhare. We know, that the fmall inteftines complete the procefs of *chylification*, which is but begun in the ftomach. I muft therefore allow, that the digeftion of animal and vegetable fubftances in the bags and tubes is perfected in the inteftines. But this is not in the leaft repugnant to the refult of thofe experiments that fhew the human ftomach to be deftitute of any triturating force, and digeftion to be the effect of the gaftric fluid alone, though the fluid which is fecreted by the fides of the fmall inteftines may complete the procefs.

ccxv. In the ccIIId paragraph I remarked, that the chief experiments on man were reducible to two heads, thofe which relate to the natural procefs, as it may be obferved by means of tubes and fuch contrivances, and thofe which relate to artificial digeftion,

digeftion, provided the gaftric juices can be procured. Having treated the former of thefe divifions as well as circumftances would permit, it remained for me to make fome enquiries relative to the fecond. It was firft neceffary to devife a method of procuring the gaftric fluid. The firft idea that ftruck me was to fearch for it in dead bodies, but after examining feveral ftomachs I was obliged to abandon this fearch; for they were either without any fluid, or elfe what they contained was fo turbid and fo much adulterated with heterogeneous matters, that it would by no means fuit my purpofe. Nor were the little fpunges, which had ferved fo well in animals, better adapted to the prefent occafion. Two fpunges would not fupply me with a fufficient quantity, and I could venture only to fwallow two tubes at once, for fear of forming an obftruction in my ftomach. Befides, the juice thus procured would have been very impure, on account of the heterogeneous matters that the tubes muft neceffarily have imbibed during their paffage through the inteftines.

There remained only to obtain it by exciting vomiting while the ftomach was empty. To effect this, I chofe rather to tickle the fauces than drink warm water, as in this cafe

the

the gaftric fluid muft have been diluted. In this manner therefore, before I took meat or drink, I procured in two mornings a quantity fufficient for a few experiments, of which the refult fhall be related below. I could have wifhed to have made a greater number, but the difagreeable feelings occafioned by the act of vomiting, the convulfions of my whole frame, and more efpecially of my ftomach, that continued for feveral hours after it, left upon my mind fuch a repugnance for the operation, that I was abfolutely incapable of repeating it, notwithftanding my earneft defire of procuring more gaftric liquor.

CCXVI. I was therefore obliged to content myfelf with what thefe two vomits afforded me. The firft time it amounted to an ounce and thirty-two grains. It was frothy at its being thrown up, and fomewhat glutinous. After it had been at reft a few hours and depofited a fmall fediment, it was as limped as water. It was a little falt to the tafte, but not at all bitter. It did not, either when thrown on the fire or brought near a candle, fhew any token of inflammability (*a*). It evaporated

---

(*a*) From this and the LXXXIft, CXXIIId, CXLIXth, and CLXXXVth paragraphs we may collect, that the gaftric juices both of man and animals are deftitute of inflammability. I
made

evaporated in the open air, and when I put fifty-two grains into a veffel and fet it on hot coals, it emitted a thick fmoke. Another fmall portion, weighing eighty-three grains, was put in a phial, which was clofed with a ftopple to prevent it from evaporating. It did not change colour or tafte, nor did it acquire any bad fmell, notwithftanding it was kept above a month in the hotteft feafon of the year. I thus employed about one half, the remainder was ufed for an attempt to obtain artificial digeftion. It was put into a glafs tube two inches long, fealed hermetically at one end, and very narrow at the other; I then introduced a fmall quantity of mafti-

---

made thefe experiments, becaufe Reaumur thought that that of his kite was inflammable, which quality Dr. Batigne imputes to the bile, a fluid confifting principally of oil *(premiere Reflexion fur les Exper. de Reaumur)*. But were this true, the gaftric juices of moft of my birds ought to have taken fire. As all mine are contrary to Reaumur's fingle experiment, I fhould fufpect, that what he obferved was owing to accident. His experiment was the following : To take away the fmell of 'putrid flefh, which one of his tubes had acquired, he fet it upon fome burning coals, when immediately there iffued a flame from the infide that lafted above a minute (Seconde Mem.). But it is eafy to perceive, that this might have been owing to fome fat of the enclofed flefh adhering to the tube. I am more confirmed in this fufpicion from having obferved, that the gaftric fluid of a kite, fuch as Reaumur's, mentioned in a note to paragraph CLXXV, was not more difpofed to take fire than the other gaftric juices which I examined.

cated

cated boiled beef, and ftopping the tube with cotton, fet it in a ftove clofe to a kitchen fire, where there was a confiderable heat, though not perhaps exactly equal to the temperature of my ftomach. By the fide of this tube I placed another, containing the fame quantity of flefh immerfed in water. The appearances in both were the following: In twelve hours the flefh in the former began to lofe its fibrous ftructure, and in thirty-five it had fo far loft its confiftence, that when I attempted to lay hold of it, it flipped from between my fingers. But though to the naked eye it appeared to be reduced to a pultaceous mafs and to have loft its fibrous texture, yet the microfcope rendered the fibres vifible; they were however reduced to a great degree of minutenefs. After this femifluid fhapelefs mafs had continued two days longer in the gaftric fluid, the folution did not feem to have made any further progrefs, and the reduced fibres were ftill juft as apparent. The flefh did not emit the leaft bad fmell, while that immerfed in water became putrid in fixteen hours, and grew worfe and worfe the two following days. It loft in fome meafure its fibrous ftructure, as always happens during putrefaction; but this appearance did not proceed

proceed fo far as in the other portion, for the fibres were entire on the third day.

CCXVII. I vomited the fecond time more gaftric fluid, and was now enabled to examine it again as I had done before; and it appeared to poffefs exactly the fame properties. In order to determine the influence of heat two tubes were filled with it, and fome flefh was immerfed as before (CCXVI). One of the tubes was placed in the ftove, and the other left in the open air. In the former the flefh was juft as much diffolved as in the preceding experiment; but in the latter the folution proceeded no farther than when water was employed (CCXVI). There was however no putrid fmell, though the flefh was left immerfed in the gaftric fluid feven days.

Before I conclude this account, I muft mention a circumftance that happened the fecond time I procured gaftric liquor by vomiting. Four hours before I fubmitted to this difagreeable operation, I had fwallowed two tubes filled with beef, one of which was thrown up; the flefh was thoroughly foaked in the fluid of the ftomach, and the furface was foft and gelatinous; it had moreover wafted from fifty-three to thirty-eight grains. This experiment proves, that there takes place a confiderable degree of digeftion in the ftomach,

ftomach, before the food paffes into the inteftines.

CCXVIII. We may now fafely lay down fome general confequences concerning digeftion in Man and animals. In the experiments on birds with mufcular ftomachs, we have feen how trituration difpofes the food to be digefted. Hence Nature has furnifhed that clafs with gaftric mufcles of fufficient power to effect this neceffary preparation. But we have likewife feen how digeftion, which confifts in the tranfmutation of the aliment into chyme, is the effect of the juices alone with which the ftomach abounds (Diff. 1).

We next proceeded to birds with intermediate ftomachs, fuch as crows and herons, and found, that in them digeftion was owing to the gaftric fluid alone (Diff. 11).

We next confidered animals with membranous ftomachs, a clafs fo numerous and various, that it comprehends almoft every family of living creatures; it includes the inhabitants of falt and frefh water; amphibious animals, as the frog, the newt, and water-fnake; reptiles, as the viper, the landfnake, and many others; quadrupeds, as the cat, the dog, the horfe, the ox; birds, as

birds

birds of prey: to this catalogue Man himself is also to be added.

In several of these animals we have seen the necessity of previous trituration, as in the ruminating order and in Man; in them it is produced by the teeth, as in gallinaceous fowls by the muscles of the stomach. But in others, as in the frog, the newt, serpents, and birds of prey, it has no share in the process of digestion. But in the latter, as well as the former cases, we have seen how the food is dissolved and digested by the gastric fluid (Diss. III, IV).

In every order of animals, Nature, always uniform in her operations, employs one principle for the performance of this vital function. Hence she has so copiously furnished the œsophagus and stomach with glands, follicles, and other contrivances that answer the same end, whence continually flow the juices so necessary to the life of Man and animals. These juices agree in many properties, but the difference of effect shews, that they differ in others. In the frog, the newt, scaly fishes, and other cold animals, the gastric fluid produces digestion in a temperature nearly equal to that of the atmosphere. But the gastric fluid of hot animals is incapable of dissolving the aliment in a degree of

heat

heat lower than that of the animals themselves. There is also a difference in celerity of action, and in efficacy. In celerity, because the food in hot animals is digested in a few hours; whereas, in the opposite kind it requires several days and even weeks, particularly in serpents. In efficacy, because the gastric juices of some animals, as the gallinaceous class, can only dissolve bodies of a soft and yielding texture, and such as have been previously triturated; while those of others, as serpents, the heron, birds of prey, the dog, decompose substances of great tenacity, as ligament and tendon, and of considerable hardness, as the hardest bone. Man belongs to this division; but his gastric fluid seems to have no action on the hardest kind of bones. Further, some species, as birds of prey, are incapable of digesting vegetables. But Man, the dog, the cat, crows, &c. dissolve the individuals of both kingdoms alike. In general these juices produce their effects out of the body, as the numerous instances of incipient digestion under this circumstance, both with the gastric fluid of animals and Man abundantly shew.

CCXIX. Having thus brought into one point of view the principal circumstances relative to the efficient cause of digestion, let

us

us compare them with what has been most plausibly written upon this topic so interesting to the physician. The opinion that prevails chiefly in the schools of Europe is that advanced by Boerhaave, who has in truth done nothing but reconcile the opinions that had been proposed at different times before him. He observes, in the first place, that the various solid and fluid substances which serve for food, being received into a close, moist, and warm vessel, must, according to the nature of each, sooner or later begin to ferment or putrefy. There are also various fluids continually running into the cavity of the stomach, viz. the saliva, the œsophageal liquor, that thin transparent fluid which is secreted by the gastric arteries, and a viscid humour secreted by glands in the stomach. If we consider the properties of these ingredients, and moreover take into the account the remains of the food which serve as a ferment, the air which produces an intestine movement of the integrant parts, the heat which excites this heterogeneous mass, we shall find, that the aliment will be macerated, diluted, attenuated, dissolved, determined to an incipient fermentation, and in short, impressed with the primary principle of vitality. Thus it is that Boerhaave explains the digestion of soft food. With respect to that of a firmer texture, imagining, that the causes above recited are insufficient to explain the digestion of them, he has recourse to the triturating power of the stomach, produced by the action of the muscular coat, and the pulsation of the aorta and the other adjacent arteries; the nervous fluid, which perhaps flows into the stomach more copiously than elsewhere; and lastly, the continual and strong compression of the diaphragm and abdominal muscles. In consequence of these additional causes, the food in the first place, will be broken down into a pulp, and acquire a cineritious hue; secondly, the fibres, tendons, cartilages, &c. will be deprived of their juices while they retain their cohesion; thirdly, from vegetable and animal substances thus dissolved, will be produced a fluid resembling those of the human body.

ccxx. Thus has this celebrated physician explained his ideas concerning digestion in his Institutions. He supposes, that there are two principal agents in this vital function, viz. the different fluids that are collected in the stomach, and the mechanical action of that organ; the secondary agents are heat, air, the nervous fluid, the remains of the food, and an incipient fermentation.

With

With refpect to the gaftric fluid, his ideas were indeterminate and unfettled. On comparing this paffage with his Prælections it will appear, that he conceived that it acted in the folution of the food like a fimple diluent, as water heated to the fame degree. But facts without number related above fhew, that it does not act in this manner, but as a real folvent. That the folution is more fpeedy and effectual than that obtained by mere water, appears from experiments equally numerous. Moreover, this fluid does not diffolve foft and yielding fubftances only, but the hardeft and moft tenacious, contrary to Boerhaave's opinion.

With refpect to trituration, the attentive reader will eafily anticipate my anfwer. However remarkable the effects produced by the mechanical action of mufcular ftomachs may be, intermediate and membranous ftomachs have no fuch power. I have made particular obfervations on the ftomach of the dog, which fo nearly refembles that of Man, and it never appeared to have any motion fufficient to break down the food. This was not only proved by thin tubes receiving no injury, but by infpection of the ftomach during the time of digeftion (CXCIX, CC). The reader will find fimilar proofs taken from the effects pro-

duced

duced by my own ftomach, in the ccixth and ccxth paragraphs. Thefe direct arguments fhew the infufficiency of the Boerhaavian hypothefis. It is befides eafy to fhew its falfity, by examining the foundation on which it refts. He deduces the triturating power from the action of the mufcular coat and the contiguous parts; but this coat is fo thin in membranous ftomachs, that its effects muft needs be inconfiderable. Nor is the preffure of the adjacent parts of much importance, at leaft in the cat and dog; for upon opening the abdomen and feeling the ftomach, I perceived nothing but the pulfation of the arteries, as I had before done in fome birds with mufcular ftomachs (xxxviii). But this pulfation does not comprefs the ftomach. I likewife perceived by my touch, that this vifcus is affected by the vibrations of the neighbouring arteries; but the effects of thefe vibrations are not more confiderable than the pulfation of its own arteries. The whole ftomach was lifted up, and depreffed by the motion attending refpiration. The periftaltic movement was alfo general in fome cafes; but the former did not produce contraction, and the contraction produced by the latter was gentle, and incapable of triturating the aliment.

aliment. It could only agitate, and so dispose it to be more readily dissolved.

CCXXI. Heat, I readily agree with Boerhaave, in considering as a co-operating cause. My experiments prove its great importance. Though the gastric fluid is not inflammable (LXXXI, CXXIII, CXLIX, CLXXXV, CCXVI), yet it is disposed by warmth to insinuate itself into digestible substances, and reduce them to that gelatinous matter which serves immediately for nutriment. The same observation is applicable to menstrua in general.

I willingly admit, that particles of air, while they are extricated from the food among which they are entangled by means of the saliva, contribute to its more speedy solution.

But I cannot so readily allow, that digestion is promoted by the nervous fluid flowing copiously into the stomach; for its very existence is uncertain, and the hypothesis is altogether without foundation.

Much less can I grant, that the remains of the aliment serve any such purpose as he ascribes to them. The great Haller justly observes, that our appetite and digestion are good only when the stomach is empty (*a*). I have had several opportunities of seeing this

(*a*) Phys. T. 6.

confirmed.

confirmed. When I fed a crow, a heron, or a falcon sparingly, the stomach would be empty in six or seven hours; when they would take food again very greedily and digest it completely, as I found upon opening the stomach.

Whether an incipient fermentation contributes to digestion, according to the opinion of this writer, is a question which shall be examined at some length in the following dissertation, as it has been the subject of many modern experiments.

Lastly, I must again differ from him with respect to fibres of flesh, membrane, tendon, cartilage, bone, which, in his opinion, are not digested in the human stomach, but only have their juices expressed; for the experiments I made on myself prove, that the solid parts are really dissolved, if we except only the hardest bones (CCV, CCVIII, CCXII, CCXIII, CCXIV). As Boerhaave endeavoured to reconcile the various opinions of physicians concerning digestion, he seems inclined to adopt in some measure the notion of those who suppose, that the office of the stomach consists in extracting the juice of animal and vegetable matters, among whom Mr. Hecquet has particularly distinguished himself. And a note to the passage, in which he observes, ferves, that the ftalks of hay are ftill vifible in the dung of the horfe and the ox, notwithftanding it is chewed fo often by the latter, ftill more clearly explains his idea. I confidered it as of great importance to enquire, whether the fame thing happens in animals belonging to òther claffes, which was really the cafe in fome. We have feen, that the two fpecies of crow above-mentioned are both granivorous and carnivorous. I fometimes fed them with wheat a little bruifed, and notwithftanding they feemed to eat it greedily, their excrements confifted of dry fragments of this grain. This likewife happened when they had eaten tough flefh. If I put the excrement in water and fhook it brifkly, the greater part would be fufpended, but a little would fall to the bottom; this, upon examination, proved to be the cellular fubftance with a few mufcular fibres, of which the particles cohered pretty firmly; the longeft pieces meafured about an inch. What remained fufpended in the water was more than twice as much as that which fell to the bottom, and ftill retained the characters of flefh. Young crows, which digeft more fpeedily than the adult (LXIX), do not completely digeft tough meat. I could eafily find cellular fubftance among their excrement; but when

inftead

inftead of hard they were fed with tender flefh, and with fome foft vegetable inftead of wheat, the excrement did not fhew the leaft appearance of this fort.

CCXXII. I made the fame obfervation upon frogs. As thefe animals generally feed upon infects, I often found among the excrement, when treated in the way juft defcribed, legs, thighs, and wings of locufts, and the cruftaceous parts of other infects.

Leuwenhoeck, upon examining the excrement of the melvel, found, that it confifted of filaments refembling the hairs of the beard cut off by the razor; thefe filaments he fuppofed to be the undigefted remains of the fifhes which the melvel had eaten (*a*). I can eafily believe this to have been the cafe, efpecially as it coincides with an obfervation of my own upon the excrement of the tench, in which, though I could not perceive any flefhy fibres, yet the fragments of bone were diftinctly vifible. I muft however add, that though I have examined the fœculent matter of many other fifhes with glaffes of various magnifying powers, I could never diftinguifh the leaft atom that had the characters of animal or vegetable matter. I have

(*a*) Philof. Tranfact. n, 152, 1683.

obferved

observed the same thing in that of nocturnal and diurnal birds of prey. The tough flesh, of which a small part was voided along with the excrement of crows (CCXXI), is entirely digested by the eagle, falcon, and owl. This observation may be extended to a multitude of birds of various kinds, of which, though I have preserved the names in my journal, to avoid prolixity I forbear to enumerate them. Serpents, though so slow of digestion, dissolve their food so completely, that not a vestige of any organized matter appears in their excrements. This at least I have seen in vipers, water and land-snakes.

Upon comparing my observations upon excrement with those related by Boerhaave and others, I think it must be concluded, that considering animals in general, some substances of both kingdoms are voided unchanged along with their excrements, because the gastric fluid is incapable of dissolving them; but others are voided unchanged, only because they do not continue long enough in the stomach to be digested. This is fully proved by my experiments on flesh, membrane, tendon, and bone, the very substances of which Boerhaave supposed, that the solid parts were indigestible. Flesh taken spontaneously by crows, part of which

is

is voided undigested, but when kept many hours in the stomach is completely diffolved, furnishes another decisive proof of the same propofition.

It is furely not neceffary to add, that I do not wish by these strictures to leffen the high reputation of the Dutch Hippocrates. Unprovided with experiments of his own, he collected the opinions of others, and framed a syftem concerning digeftion fo ingenious and plaufible, that I willingly own that I formerly adopted it, and would not now reject it, if I was not compelled by conclufive experiments.

CCXXIII. I will conclude this differtation, with fome remarks on a problem clofely connected with refearches concerning the efficient caufe of digeftion. Mr. Hunter, one of the beft Englifh anatomifts of the prefent age, frequently found in the dead bodies which he opened that the great curvature of the stomach was confiderably eroded, and fometimes entirely diffolved. In the former cafe, the edges of the wound were as foft as half-digefted food, and the contents of the ftomach had got into the cavity of the abdomen. He obferves, that fuch a wound could not have exifted in life, as it had no connection with the difeafe, and more frequently appeared in

perfons

persons who died violent deaths. In order to discover the cause of this phænomenon, he examined the stomachs of various animals, both immediately and some time after death. In several he observed the same appearance. Hence he thought he was enabled to assign the cause. He supposes the solution to be owing to a continuance of digestion after death, and that the gastric fluid is capable of dissolving the stomach when it has lost its vital principle. From this discovery he infers, that digestion neither depends on the action of the stomach nor on heat, but on the gastric juices, which he considers as the true menstruum of the food (*a*).

CCXXIV. When Mr. Hunter's short but sensible paper came to my hands, I was engaged in experiments on digestion. I had discovered the primary importance of the gastric fluid in this process, and that it acts out of the body; that is to say, in the dead body. I knew also, that after death this fluid issues from the coats of the stomach. From these data I had little difficulty in believing the fact related by the English anatomist, and adopting his explanation of it: nevertheless it was proper to repeat the experiment. Being un-

(*a*) Ph. Transf.

provided

provided with human subjects, I had recourse to animals. Some were opened sooner, and others later after death; but among the numbers I inspected, not one had its great curvature dissolved, or much eroded. I say, much eroded, because I have often seen a little erosion, especially in different fishes, in which, when I had cleared the stomach of its contents, the internal coat was wanting. The injury was always confined to the inferior part of the stomach. If these facts are favourable to Mr. Hunter, a great number are against him. They do not however destroy his observations; mine are only negative, his are positive; and we know that a thousand of the former do not destroy a single one of the latter, provided it is well ascertained. I have no reason to distrust Mr. Hunter, for his paper has the air of ingenuousness and candour which usually accompanies truth.

ccxxv. The ill success of my experiments did not induce me to abandon the idea of digestion after death, it only led me to consider it in another point of view. If it be true, said I to myself, that the gastric fluid exerts its action after death, it must produce some solution of the food. Let then an animal be fed and immediately killed, after some time

time let it be opened, and let us fee whether the food has been at all digefted. I determined to bring this obvious inference to the teft of experiment; I therefore kept a raven fafting feven hours in order to empty its ftomach, and then fet before it an hundred and fourteen grains of beef, which were immediately eaten, and muft have paffed into the ftomach, as this bird has no crop. I then killed it, and as it was winter, put it into a ftove, where it was left fix hours. Suppofing this to be a fufficient time for the gaftric fluid to exert its action, I opened the ftomach, and found the flefh in the following ftate. It was impregnated with gaftric fluid, and was become tender; the colour was changed to a pale red, and the furface had a bitter tafte, while the internal parts retained the tafte of flefh. After the gaftric fluid was wiped away, it weighed only fifty-two grains; it had therefore loft above half its weight in fix hours, or, what amounts to the fame thing, was above half digefted. The pylorus, and the duodenum for about an inch, were occupied by an afh-coloured mucus, which muft have been the diffolved part of the flefh.

At the fame time I gave another raven, that had in like manner been kept fafting feven
hours,

hours, an equal quantity of flesh, and killed it in two hours and a quarter. My view was to observe the difference between what had lain six hours in the dead, and two and a quarter in the living stomach, and it was very great; for in this latter case the flesh was totally dissolved, except a little cellular substance, which I have found to be always longer in being digested than the muscular fibres; the mucus was the same as before, only in larger quantity, and occupied more of the duodenum. These two experiments compared together prove two things, first, that digestion continues after death; and secondly, that it is then far less considerable than in the living animal, though in the present instance the heat of the stove, which was about 100° (*a*), must have promoted it not a little. The heat of the living raven did not exceed 30° (*b*).

CCXXVI. Another dead raven was kept five hours in the same stove, after I had forced two dead lampreys, weighing together an hundred and twelve grains, down its throat. One lay in the œsophagus, the other had reached the stomach and was completely de-

---

(*a*) Two hundred fifty-seven deg. Fahr. Ther.
(*b*) One hundred nine and an half ditto.

composed,

compofed, while the former was indeed entire, but foft and flaccid. This accident proves, that the gaftric fluid is capable of producing a fenfible degree of digeftion at a time when the œfophageal juices are inert.

CCXXVII. Thefe experiments were made in winter. I determined to repeat them the next fummer, becaufe then I could expofe the dead animals to a greater heat. Accordingly in that feafon fome bruifed veal was given to two ravens, which were immediately killed, and left feven hours in a window expofed to the fun. We have already feen in feveral paffages, the influence of heat in promoting artificial digeftion (CXLII, CLXXXVI, CCI, CCXVII). Nor did it now appear lefs confiderable. Each raven had eaten fixty-eight grains of flefh, of which there was not an atom left entire; it was all diffolved into the ufual gelatinous pulp, and the greater part had paffed through the pylorus.

Thefe facts, I think, decifively prove, that animals, at leaft the fpecies juft mentioned, continue to digeft after death. If we confider the matter rigoroufly, it will be proper to obviate a difficulty that may be ftarted. However careful we are to kill the

the animal immediately after it has swallowed food, it is certain, that there will be a short interval between the time the food gets into the stomach and the death of the animal, and that the gastric fluids act upon it during this interval. Moreover, after death they will act for some time just as in life, since the vital heat is not instantly extinguished. The digestion therefore observed in dead animals may, if not entirely, at least in part, be produced by the gastric fluid acting during life, and a short time after death.

Nothing could be more easy than to ascertain the justness of this suspicion, since we have only to thrust a little food into the stomach of a dead and cold animal, and observe the consequence. I made the experiment upon a raven that had been dead an hour, and had now only the temperature of the atmosphere. Forty-two grains of beef cut into pieces were forced into the stomach, which was opened after the bird had lain seven hours exposed to the sun. And here instead of pieces of solid flesh, I found only the usual pulpy mass, partly in the stomach and partly in the duodenum. The solution was therefore effected by the gastric fluid, independently of the powers of life.

CCXXVIII.

CCXXVIII. The experiment was repeated upon an owl and a blackbird, which were killed immediately after meat had been given them, and left seven hours in a warm temperature. The flesh given to the black-bird had been cut into three pieces, which together, amounted to eighty-two grains; the owl had swallowed half an ounce and six grains in one piece. Upon opening the stomachs, I found the four pieces; but the surface was covered with a stratum of mucus, which shewed, that the flesh had been dissolved.

I thought, that perhaps if the flesh had remained a longer time in the stomach it would be more digested; but this did not happen, at least when I repeated the two preceding experiments under the same circumstances, except that the birds were exposed to the sun for twenty-two hours, I could not perceive, that the solution of the flesh was carried any further. The entrails emitted a putrid smell, but this was not the case either with the stomach or its contents.

CCXXIX. That I might be warranted in deducing general consequences, I resolved to repeat this singular experiment upon various classes of animals, and therefore had recourse to fishes and quadrupeds. Of the former, the fifh-márket at Pavia only affords the pike, carp, barb'el, tench, eel, and the like; but I took care to procure fuch as were very frefh. I introduced into the ftomach various animal fubftances, as little fifhes, bits of veal and beef, frogs, grubs, &c. and opened them after an interval, fometimes fhorter and fometimes longer. I will give in a few words what is fet down at great length in my Journals. The part of thefe fubftances that lay in the œfophagus, a pofition which they often had, was unaltered; this was fometimes the cafe with that which had got into the ftomach, but it was generally more or lefs eroded. A circumftance refpecting frogs deferves to be mentioned. The tough fkin of thefe animals was often deftroyed, efpecially at the bottom of the ftomach; and where it ftill remained it was fo much foftened, that the flighteft force was fufficient to lacerate it. Hence it appears, that the gaftric fluid of fifhes retains its property of diffolving flefh, but in an inferior degree to that of birds, fince it did not diffolve fo much.

ccxxx. The quadrupeds upon which I made thefe experiments, were dogs and cats. After keeping them fafting many hours, I gave them a certain quantity of flefh, and then

then ſtrangled them without delay. Of three dogs and three cats, two of the former and as many of the latter were expoſed to the ſun for nine hours; the others were left in the ſhade. In the firſt the ſurface of the fleſh was gelatinous as uſual, but in the laſt this appearance was ſcarce perceptible. Theſe experiments confirm the utility, I ſhould rather ſay, the neceſſity of heat to digeſtion in many animals.

CCXXXI. To conclude this curious enquiry, I reſolved to ſee what change would take place upon fleſh when the ſtomach was taken out of the body. I made this experiment upon a cat, a raven, and an owl. Having fed them ſparingly, I cut out the ſtomach, and threw ligatures round the cardia and pylorus to prevent the contents from getting out. They were expoſed to the ſun in a veſſel of water, leſt the heat ſhould dry them. In five hours and an half they were opened: the water had tranſuded through the coats; the ſurface of the fleſh was a little gelatinous, eſpecially in the ſtomach of the raven and owl; but the ſolution was trifling, in compariſon with that which took place when the ſtomach was left in the body. This was what might be expected, when the œſophagus no longer poured its liquor into the ſtomach.

In theſe experiments I did not perceive any eroſion of the ſtomach, any more than in thoſe made with the view of verifying Mr. Hunter's (ccxxiv). I only ſaw what I had ſeen before (ibid), a ſlight excoriation of the inferior part. We muſt therefore infer, that the coats of the ſtomach ſuffer leſs after death than fleſh introduced into it. I gave an hungry dog ſome pieces cut out of the ſtomach of another dog; he eat and it killed him immediately. After the body had lain in a warm ſituation nine hours, the ſtomach was opened. The pieces were ſenſibly diſſolved, but no change was produced upon the ſtomach of the animal, if we except the large curvature, which was ſo much macerated, that the villous coat might eaſily be rubbed off. It is, I think, not difficult to aſſign the reaſon, why the ſtomachs of dead animals are not liable, like their contents, to be diſſolved. Theſe bodies are inveſted on all ſides by the gaſtric fluid, whereas it acts only on the internal ſurface of the ſtomach.

Upon reviewing the experiments related in the ccxxvth and following paragraphs it cannot, I think, be doubted, that digeſtion goes

goes on for some time after death. I therefore entirely agree so far with the celebrated English anatomist, but I cannot with him suppose, that this function is independent of heat (CCXXIII); numberless facts related in this work fully prove the contrary.

DISSERTATION VI.

WHETHER THE FOOD FERMENTS IN THE STOMACH.

CCXXXII. I WILL now, agreeably to my promise in the foregoing dissertation (CCXXI), enquire whether the food ferments in the stomach. This opinion was almost universally adopted by physicians about the middle of the last century, an æra at which the explanation of the various functions of the human body was sought in fermentations of various kinds, as it had before been in a subtile matter, as it has since been in electricity, and is at present in divers sorts of elastic fluids. This notion was afterwards combated among others by Boerhaave, who found, by direct observations, that this multiplicity of fermentations did

not

not exift in nature, but was merely the fug-
geftion of fancy. Of the numberlefs modi-
fications of this procefs, which phyfiologifts
had imagined, he admitted only that very limited and imperfect one, which, according
to him, takes place in the ftomach. The
food in the ftomach of animals, and particularly of Man, is, in his opinion, in circumftances highly favourable to fermentation. The faliva and the gaftric fluid ferve
inftead of water; the free accefs of air, the
clofenefs and heat of the ftomach, the nature
of the food itfelf neceffarily produce fermentation, as is farther evident from the eructations confequent upon taking food, and
the rumbling noife frequently heard in the
belly. But the fhort continuance of the food
in that vifcus, and other caufes, prevent the
procefs from being ever carried to its utmoft
pitch.

CCXXXIII. Thus far only, according to
Boerhaave and his followers, does the fermentation of the food proceed in the ftomach. This limitation has been thought
too great by Dr. Pringle and Dr. Macbride,
two celebrated modern phyficians. They
find no difficulty in fuppofing, that a complete fermentation takes place in digeftion,
and that it is the chief agent in this important

tant function. In their researches on this subject, they have endeavoured to imitate the operations of nature out of the body. They took various animal and vegetable substances, such as are used every day for food; they placed them both by themselves, and mixed with several other substances in a warm temperature, adding a quantity of water or saliva. Under these circumstances they found, that they sooner or later began to ferment; that this process afterwards ran very high, then abated, and at last ended in the decomposition of the several substances, which acquired also a sweet taste. These different gradations of fermentation were evident from the swelling, rarefaction, and intestine movement of the mass, from the generation of a multitude of air-bubbles, and from the substances which at first sunk to the bottom, at length floating on the surface of the fluid. These experiments first made by Pringle, and afterwards repeated and varied by Macbride, determined them both to consider digestion as a process merely fermentative. Their theory is as follows. The food divided by mastication and penetrated by the saliva, begins as soon as it gets into the stomach to be agitated by that intestine movement which always accompanies fermentation; this movement

is

is excited by the warmth of the place, by the remains of food taken before, by the gaſtric fluid, and above all by the ſaliva, which is particularly adapted to produce and promote this procefs. The firſt effect of the inteſtine commotion will be to raiſe the ſolid parts of the aliment to the ſurface of the gaſtric liquor; here they will be ſuſtained for ſome time by the air-bubbles; but on their ceſſation they will fall down again and be thoroughly incorporated with the fluids of the ſtomach. The periſtaltic motion, the alternate preſſure of the diaphragm and abdominal muſcles, and the continual pulſation of the adjacent large veſſels will render this mixture ſtill more complete. In ſuch a ſtate the food paſſes into the ſmall inteſtines, where the fermentative motion produces ſtill greater changes in conſequence of the mixture of the bile and pancreatic juice. And now the various kinds of food are changed into a ſweet, mild, nutritious matter, which ferments briſkly, and is denominated *chyle*. In conformity with this theory, theſe phyſicians eſtabliſh a new ſyſtem of great importance, according to them, in the practice of medicine. It is ingeniouſly explained by Pringle, in his Appendix containing Experiments on ſceptic and antiſeptic Subſtances, and by

Macbride

Macbride in his experimental Essays on the Fermentation of alimentary Mixtures.

CCXXXIV. The opinion of these two modern writers have been adopted by many physiologists, while others have still adhered to the doctrine of Boerhaave, concerning an incipient and incomplete fermentation only taking place in the stomach; so that on this subject the physicians of Europe seem to be divided into two sects. When I read Pringle and Macbride, I had only made a few experiments on the digestion of some animal and vegetable substances enclosed in tubes by gallinaceous birds (XXXIX, XL, XLI, XLII, XLIII); and I began to perceive, that the gastric juice acted as a menstruum upon the food. But I could not learn from these experiments, whether fermentation takes place at the time they are dissolved. As indeed the gastric fluid is a solvent, it may act independently of fermentation; chemistry affords numerous instances, in which there is no token of fermentation during the dissolution of the solvend. But there is no absurdity in supposing, that an intestine fermentative motion is generated in the mixture, at the time the gastric fluid dissolves the aliment. And in this case, fermentation would accompany digestion, though it would not according to to the doctrine of Pringle and Macbride (CCXXXIII), be the efficient cause. In order to obtain information concerning this phænomenon which I had not noticed, I had recourse to further experiments. As the theory in question is entirely founded on the fermentation of animal and vegetable matters in vessels, I set in glass phials bread, flesh, and saliva; bread, flesh, and water; flour, saliva, and flesh; for in these mixtures the writers above-mentioned observed the most rapid fermentation. The phials were stopped, and set in a place where the heat amounted to 20°---24° (*a*). The mixtures began, some sooner and others later, to emit air-bubbles, which soon encreased in frequency and size; the surface of the liquor was covered with froth, which continued as long as any air was generated. During this time the mass swelled greatly, the intestine commotion was manifest, and the substances immersed being made specifically lighter by the air-bubbles that adhered to them and the increase of bulk, rose to the surface of the fluid. Here then the tokens of fermentation were apparent, and so far I entirely agree with Pringle and Macbride.

(*a*) Seventy-seven and eighty-six deg. of Fahr. Ther.

CCXXXV. But

CCXXXV. But found logic forbade me to allow fo readily, that the fame procefs takes place in the ftomach. I had indeed many reafons for with-holding my affent. Not to mention the fhort continuance of the food in that vifcus, a circumftance which did not efcape Boerhaave (CCXXXII). I confidered, that although the faliva produces and promotes fermentation, the gaftric fluid may not have this property. Though the gaftric fluid confifts in part of faliva, yet as there are feveral other ingredients, a compound muft be formed with properties different from thofe of its conftituent parts. I have adduced many inftances to prove, that the gaftric fluid retains in fome meafure its folvent power out of the body; but the faliva never exhibited any fuch property. I have already fhewn, and fhall ftill more clearly fhew in the fequel, that flefh immerfed in the gaftric fluid is not liable to putrefaction; but when put into faliva, it putrefies fooner than in water. This was one of my motives for not immediately adopting the ideas of Pringle and Macbride. It were to be wifhed, that thefe phyficians had made trial of the gaftric liquor alfo, before they concluded, that what they obferved in veffels takes place likewife in the ftomach; nor can I well conceive, how they

both

both came to overlook a circumstance of so great importance. Moreover we know, that rest is necessary to fermentation; but the stomach, besides the motion of the whole body, has a movement peculiar to itself. Lastly, should fermentation once begin, it must in all likelihood be soon stopped by the fresh saliva and gastric liquor that are running perpetually, and in no small quantity, into the stomach. These two last objections have been already started, though nobody, as far as I know, has taken the pains of verifying them by experiment. But as the question could be decided in this way only, I determined to undertake to supply the omission.

CCXXXVI. I have already spoken of artificial digestion in several passages. Experiments of this kind afforded me an excellent opportunity of observing, whether the solution of flesh out of the body was accompanied by fermentation, and I never failed in a single instance to attend to this circumstance. I found, that when the vessels remained at rest, a few small air-bubbles began to arise in the space of a few hours; they afterwards became larger and more frequent, and adhering to the immersed substances, caused them to rise to the surface of the liquor. This air was either entangled in the mixture, or, according

cording to Pringle and Macbride, formed part of it, was extricated, and rendered elaſtic by the heat, or what ſeems more probable, came from both theſe ſources. The mixtures either ſunk again or continued to float, while they were diſſolved by the gaſtric menſtruum; not the ſlighteſt inteſtine motion was ever perceptible, juſt contrary to what happens when ſaliva is employed. If I now and then ſhook the veſſel a few hours after making the infuſion, very few air-bubbles were generated, and the mixture hardly ever roſe to the ſurface, though it was juſt as well diſſolved as when the veſſel remained at perfect reſt. I find in my Journals, that I agitated the veſſels upon fourteen different occaſions without obſerving the ſmalleſt difference in the reſult of the experiment. I could not therefore allow, that fermentation was the efficient cauſe of theſe artificial digeſtions, nor even that it was a concomitant circumſtance, or an effect; and freſh experiments inclined me more and more to reject this opinion. I have already mentioned the great abundance of gaſtric fluid in crows, and the facility with which they digeſt their food, more eſpecially neſtlings (LXIX, LXXXIII). Among the various trials I made with this fluid out of the body, I endeavoured

deavoured to renew it, as it is renewed in the ftomach. Several glafs tubes were filled with it to a certain height, and fufpended in a vertical pofition; into the upper extremity a fmall funnel was put; fome gaftric fluid was poured into it from time to time, the narrownefs of the orifice of the funnel allowed it to fall only drop by drop into the tubes. The lower extremity of the tubes was not clofely ftopped, that nearly as much might run out below as fell in from above. Matters being thus arranged, I immerfed in the tube flefh and bread, both by themfelves and mixed together. The folution was exceedingly fpeedy, on account both of the warmth of the atmofphere, and the conftant renewal of the gaftric fluid. Notwithftanding the tubes remained at perfect reft, only a few air-bubbles were difcharged; not the leaft inteftine motion could be perceived; the flefh and bread fell immediately to the bottom, and remained there till they were gradually incorporated with the gaftric fluid: in fhort, they were digefted without a fingle circumftance occurring that ufually attends fermentation.

CCXXXVII. If this procefs does not take place out of the body, it feems highly improbable that it fhould within; however, to be

certain

certain of this, it was proper to confult the fenfes. Is digeftion, according to Pringle and Macbride, a fermentative procefs? Let us then obferve it while it is going on, furprize Nature in her operation, and fee in what it confifts. With this view, I gave four hens that had been kept fafting twelve hours, fome wheat, and in five hours opened the gizzard without killing them. This method I practifed in the following experiments, being apprehenfive left opening the animal after death might not anfwer the end I had in view. Both gizzards were full of grains of wheat moftly broken, and mixed with a femifluid farinaceous pafte. The orifice of the pylorus and great part of the duodenum was full of the fame pafte, which had not in this cafe much fluidity. Upon examining this pafte, both with my naked eye and the microfcope, I could not perceive any fign of fermentation; the parts were at perfect reft, and entirely free from air-bubbles. I waited three hours longer before I opened the gizzards of the other two ducks, in order to fee whether what had not taken place at the beginning of the procefs, might not have happened when it was further advanced. In this cafe, the pafte was more diluted with gaftric liquor, and of the grains of wheat little was left but the

bran;

bran; I observed no more inteftine motion or air-bubbles than before.

CCXXXVIII. My next experiment was made upon three ravens that had not yet quitted the neft. Two hours after I had fed them with beef, I opened the ftomach of one of them. The flefh was half diffolved, but I could not perceive any fign of fermentation. I made the fame obfervation upon the two others, which were opened an hour and three quarters afterwards, notwithftanding digeftion was finifhed; for nothing remained in the ftomach but a denfe grey fluid, confifting of flefh diffolved in the gaftric fluid.

Of animals with membranous ftomachs I examined an owl, feveral dogs, cats, and land and water-fnakes, endeavouring always to make my obfervations at three diftinct times, at the beginning, towards the middle, and at the end of digeftion. But at no time did I perceive any tendency to fermentation. In one dog and one cat only did I obferve a few air-bubbles among the food after it was completely digefted; but there was not the leaft inteftine motion perceptible. Serpents, which are animals fo flow of digeftion, were well adapted to fhew the progrefs of this function; but neither did they form an exception to the general obfervation. Thefe facts

facts obliged me to reject the opinion of the British physicians and their followers; nor do I know whether that of Boerhaave is admissible, who, while he excludes a complete, infers an incipient fermentation from the eructations that arise after taking food (CCXXXII); but this may be occasioned by the rarefaction of the air entangled among the aliment, by the mere heat of the stomach.

CCXXXIX. Modern chemists have distinguished three species or degrees of fermentation, the vinous or the sweet, the acetous, and the putrid. As they essentially consist in an intestine motion excited by heat and a proper degree of moisture (*a*), and as no such motion can be seen in the food in the stomach, it follows, that not even the vinous, much less the acetous or putrid takes place in digestion. It remains to be enquired, whether this function is connected with a principle of acidity, as some suppose, or of putrefaction, according to others. I shall state the facts which appear to favour each of these opinions. In behalf of the first, its advocates adduce acid eructations and vomitings from the human stomach, that disagreeable

---

(*a*) Macquer Dict. Art. *Fermentation*.

acid

acid smell which is exhaled from the stomachs of granivorous birds and ruminating animals, the acetous taste of the internal coat; the diminution of the bulk of the contents of the stomach of man and animals, not to mention other arguments that may be seen in modern physiologists, and especially in Haller.

CCXL. The prodigious number of stomachs I have opened, have afforded me opportunities of acquiring full information on this point. In animals strictly carnivorous, such as birds of prey and serpents, the food never has an acetous taste, either to the taste or smell during the time of digestion. The same observation will apply to frogs and fishes; and it may be extended to omnivorous animals, such as crows, when they feed upon flesh; but the pultaceous mass resulting from vegetables, and in particular from bread, now and then acquire a slight acidity. I have observed the same taste in two dogs, and more frequently in herbivorous animals, such as sheep and oxen; and also in those which are at once herbivorous and granivorous, viz. in the gallinaceous kind; and in the last-mentioned class, not only the food in the stomach had an acetous taste, but that in the craw likewise. In the third dissertation will be found some instances of this (CXXXIX,

CXL,

CXL, CXLI, CXLIII). With refpect to Man, I will relate what has happened to myfelf. During the whole month of May and great part of June, I eat ftrawberries with fugar and white wine at dinner and fupper. From this agreeable mixture I never experience any inconvenience in the day-time; but by what I eat in the evening my fleep is frequently difturbed; the contents of my ftomach rife almoft into my mouth, and then fall back again, leaving a moft difagreeable four tafte behind. This unpleafant circumftance does not, however, prevent me from recovering my reft, and digefting my food perfectly. I have befides many times been fubject to a like difagreeable fenfation, after eating too much fruit in fummer and autumn. Every man muft fome time or other have been fenfible of his meat and drink having turned four.

CCXLI. I was further defirous of knowing, whether the acid principle fometimes found in the ftomach, is capable of diffolving calcareous earth, and fuch other bodies as acids act upon. I accordingly gave fome carnivorous birds pieces of coral and fea-fhells, and they were thrown up without any change of colour, or diminution of weight. This was what might be expected. I next made
the

the fame experiment on a hen and a turkey, which were killed in two days. The coral and fhells were very much corroded, and the former was reduced to pieces; but a moment's reflection fhewed me, that the corrofion might be owing to the action of the ftomach, and not to an acid. The doubt, however, might eafily be removed, by enclofing the fame fubftances in ftrong metallic tubes. The refult of feveral experiments made in this manner was, firft, that the pieces of coral and fhell were almoft always diminifhed; but the diminution fcarce ever exceeded three or four grains: fecondly, that the furface was foftened; and thirdly, that it was turned black, efpecially in the coral. I immerfed at the fame time the fame fubftances in diluted vinegar, and as fimilar effects were produced, and particularly the black colour, I inferred, that the phænomena rofe from a like caufe; laftly, I repeated the experiment on myfelf. The tubes were covered as before (CCVIII), to prevent any feculent matter from getting into them. They were all voided without inconvenience. When I eat flefh, with the addition only of a little bread, the fubftances were neither diminifhed nor altered in their colour. But upon eating a large quantity of different vegetables, the coral and

and shells were generally diminished and darkened. These facts prove the presence of an acid principle in the stomachs of some animals, and Man himself. It is, however, not perpetual, but depends on the quality of the food.

CCXLII. This acid soon disappears. I gave several gallinaceous fowls some bread at the same time. The stomachs were examined at different intervals, viz. two, three, three and an half, four, and five hours afterwards. As long as the bread preserved its consistence, it was frequently acid; but as soon as it was reduced to chyme this taste was totally lost. Nor could I perceive the least sign of it in that which had passed into the duodenum. I made upon myself the following observation.

When the unpleasant acid taste mentioned above (CCXL) came into my mouth in consequence of eating strawberries, I kept myself awake twice during the remainder of the night. Acid eructations continued to arise for some time; they at last ceased; yet from a sense of weight I knew, that the contents of the stomach were not entirely digested; but the flatulence that came from them had no longer the slightest acidity.

CCXLIII. But

CCXLIII. But what produces this acidity in the ſtomach? Does it come from the gaſtric fluid, or from the food? There are good grounds for rejecting the former, and admitting the latter of theſe ſources. In the firſt place, this acidity does not appear on all occaſions; I never could obſerve it ariſing from fleſh. Now if it came from the gaſtric fluid, why ſhould it not be communicated to every kind of food, ſince every kind is alike impregnated with it? Secondly, when I eat vegetables, the effects of an acid in my ſtomach were apparent, but not when I eat fleſh (CCXLI). Thirdly, when vegetable food is completely diſſolved by the gaſtric fluid, it then loſes all acidity (CCXLII). Laſtly, if acid bread be encloſed in tubes and given to a crow, when it is thrown up four or five hours afterwards, the little that remains inſtead of being ſour, is now turned ſweet.

CCXLIV. Notwithſtanding theſe proofs, that this acidity is not owing to the gaſtric fluid, but to the tendency the food itſelf has to turn ſour whenever it is in a warm temperature, is it not ſuppoſed, that this fluid both in Man and animals is of an acid nature? Have not moſt of the antient and many modern phyſicians ſubſcribed to this opinion? I ſhould therefore have incurred the reproach of negligence, if I had not undertaken a chemical analyſis of it. The gaſtric fluid of every animal mentioned in theſe diſſertations, not excepting my own, was ſubmitted to the following experiments. Having taken the precautions above deſcribed (LXXXI, CCXV), to procure it in a ſtate of purity, I dropped it upon ſalt of tartar *per deliquium*, and into the nitrous and marine acids, without ever perceiving any change of colour, any motion or efferveſcence; whence I was obliged to infer, that the gaſtric fluid is neither acid nor alkaline, but neutral. I thought it would alſo be proper to ſubject thoſe kinds which could be procured in large quantity, as that of the crow, to the action of fire; I therefore entreated my illuſtrious colleague and friend, Counſellor Scopoli, to undertake the analyſis, as he was not only provided with the proper apparatus, but eminent for his ſkill in chemiſtry, of which ſcience he is deſervedly public profeſſor. He complied with my requeſt, and in a few days favoured me with the following account.

Chemical Analysis of the Gastric Fluid of the Crow.

" The liquor is turbid, and of a darkish colour. When shaken it emits a smell rather disagreeable.

When triturated with quick-lime or salt of tartar, a fetid urinous odour is exhaled.

It does not effervesce with either of the mineral acids. It gives rather a green hue to syrup of violets.

Two drachms exposed to a gentle heat left a dark-coloured sediment weighing two grains, which attracted the humidity of the air. This residuum had a nauseous smell. It did not effervesce with acids.

I next filtered and distilled it. A darkish matter was left upon the filter, which, when it was dried, appeared in the form of a nut-brown powder, of a salt and bitter taste.

The liquor which passed into the receiver was divided into five portions. The first had a slight taste, and an empyreumatic smell. The second had a stronger taste and smell. The third, fourth, and fifth resembled the second, but the last had the strongest empyreuma.

The

The belly of the retort was almoſt entirely covered with a white ſaline ſubſtance, which upon being triturated with quick-lime emitted a fetid urinous ſmell. In the bottom there remained a tough dark-coloured ſubſtance, reſembling an extract. It did not effervefce with acids; its ſmell was empyreumatic, and its taſte ſalt, bitter, and nauſeous. This ſalt is neither acid nor alkaline, for it does not effervefce either with acid or alkalies; but when a little oil of tartar *per deliquium* is mixed with it, it emits a penetrating urinous odour, exactly like that of ſal ammoniac.

From theſe experiments we may conclude, that the gaſtric fluid contains, firſt, pure water; ſecondly, a ſaponaceous and gelatinous animal ſubſtance; thirdly, ſal ammoniac; fourthly, an earthy matter like that which exiſts in all animal fluids.

The ſaponaceous ſubſtance altered by fire emits that unpleaſant empyreumatic ſmell.

The ſal ammoniac being enveloped by the ſoapy matter does not ſublime, as it does when not entangled by other ſubſtances.

The gaſtric fluid of the crow precipitates ſilver from nitrous acid, and forms luna cornea. This phænomenon might induce us to ſuppoſe, that common ſalt exiſts in the gaſtric

tric fluid; but as the falt contained in this fluid is not common falt but fal ammoniac, we muft fuppofe, that the filver is feparated from the nitrous, on account of its ftronger attraction for the marine acid, which alfo far exceeds the attraction of the volatile alkali for the latter acid.

I wifh you would repeat thefe obfervations on the gaftric fluid of animals feeding only on vegetables. If in this alfo fal ammoniac fhould be found, we muft conclude, that the marine acid is generated by the animal powers; and we might fufpect, that the marine acid of fea falt is produced by the animals that inhabit the ocean. This is however a mere conjecture.

I am, &c.

Scopoli."

A little after I had received this account from my celebrated colleague, I quitted Pavia, to fpend the fummer vacation in my own country, where I had no opportunity of making experiments on the gaftric fluid of any animal ftrictly herbivorous, though I earneftly wifhed for it. I obtained, however, fatisfactory

factory proofs from the raven, that the ammoniacal salt does not depend on animal food, but on the powers of life. I fed five ravens for fifteen days on vegetables alone, and then by means of spunges procured a quantity of gastric fluid, which I supposed would have no properties that could be ascribed to animal food. When I made with it the experiments described above, it did not appear to be acid or alkaline; it had a salt taste, and upon pouring a few drops into a solution of silver in the nitrous acid, luna cornea was precipitated. There is therefore every reason to suppose, that if this fluid was distilled, sal ammoniac would be obtained; and therefore, that the marine acid is the product of the animal powers. But whatever we are to think either of this or the other ingenious conjecture of my colleague, which have indeed little connection with our present enquiry, it is certain, both from his experiments and my own, that the gastric fluid is not either acid or alkaline, but neutral.

CCXLV. But I must not conceal those arguments which are adduced to prove, that there is a latent acid in this fluid, though it cannot be detected by any of the ordinary chemical means. It is well known, that a small quantity of acid will curdle milk, an effect

effect produced in the stomach of animals, and of sucking calves in particular, in which case we cannot suspect any vegetable acid to be present; the phænomenon must, therefore, be attributed to the latent acidity of the gastric fluid. And as it is continually secreted by the internal coat of the stomach, we cannot be surprized, that this coat in some animals should retain the property of curdling milk. This is well known to cooks, who, when they have no rennet, take the innermost coat of the stomach of a fowl and steep it in water; which water, when thus impregnated with the juice of the stomach, will serve for turning milk as well as rennet itself.

Hence some have supposed, that the stomachs contains a latent acid. My first step was to ascertain the fact. I therefore triturated the internal coat of a hen with water, which was thus rendered turbid, and in an hour and a half curdled a quantity of milk. The same effect was produced by the internal coat of other gallinaceous birds, viz. the capon, turkey, duck, goose, pigeon, partridge, quail, treated in the same manner. I further discovered, that this property belongs also to intermediate and membranous stomachs, by experiments on that of the crow, heron, birds of prey, the rabbit, the dog, cat, various

reptiles,

reptiles, and several scaly fishes. In these trials the stomachs were fresh. I next tried dried ones, chiefly taken from the gallinaceous class, which being almost of the consistence of horn, become dry in a very short space, and at the same time exceedingly brittle. The results were the same as before. Nor did it make any difference, though they had been kept ever so long. I have had for three years the internal coats of the stomach of several fowls, and upon triturating them with water, while I am writing, they curdle milk as well as at first. If they are pounded and mixed with milk, they answer the purpose equally well.

CCXLVI. But is this property confined to the internal coat? It was easy to determine this by treating the others in the same manner. The nervous coat has this property in some degree, but falls fart short of the internal. Whether cut into small pieces and macerated in water, or mixed immediately with milk, the effect is not so speedily produced, nor so considerable, nor are the curds so hard. The muscular and cellular coats have not this property in the smallest degree, at least in gallinaceous birds, upon which these experiments were made. Hence it would seem, that it resides in the internal coat solely; for
the

the effects produced by the nervous coat, may be owing to its lying in contact with the former.

CCXLVII. But is this property inherent in the internal coat, or adventitious, and owing to the gaftric fluid with which it is impregnated? I incline to the latter opinion, since the gaftric fluid fo readily curdles milk. I fhould weary my reader, was I to recount all my experiments. I will therefore only fay, that the gaftric fluid, from whatever animal it was obtained, poffeffes this property, whether procured by fpunges, by opening the ftomach and expreffing it out of the glands, and the mouths of the little arteries, with which this vifcus in general abounds. 1 have further found, that the gaftric fluid need not be frefh. That of crows, at leaft, preferves its virtue for three months.

CCXLVIII. But is it a neceffary confequence of thefe experiments, that the gaftric fluid contains an acid? As no chemical teft fhews this quality, there can be no juft motive to admit it, unlefs it can be proved to be a neceffary confequence of the curdling of milk. This is maintained by the illuftrious Macquer among others, who is of opinion, that whatever bodies of the animal and vegetable

kingdom

kingdom coagulate milk, have either a manifest or occult acidity (*a*).

The foundation of this opinion is the common obfervation, that acids are the fole caufe of the curdling of milk. To this reafoning, I fhall only oppofe a fingle fact: I have difcovered, that though feveral animal fubftances are incapable of producing this effect, yet others have this property. Thus for inftance, if the blood or bile of a turkey be mixed with milk, it will retain its fluidity; but pieces of the heart, liver, or lungs of that bird, will curdle it readily. This obfervation is not merely owing to accident; I have made the experiment repeatedly with different turkeys, and always with the fame fuccefs. If therefore the coagulation of milk be always owing to an acid, we muft fuppofe an acid in the heart, liver, and lungs of the turkey. I am aware, that many chemifts, in oppofition to the Boerhavian fchool, think a real acid exifts in the different parts of animals, and particularly in the blood; but according to this hypothefis, I cannot comprehend, why the blood of the turkey and other animals does not coagulate milk. With refpect to the latent acid of the gaftric fluid, I fhall

(*a*) Art. Milk.

very willingly leave my readers to adopt what opinion they shall think most probable. The milk I employed for my experiments, was sometimes that of the sheep, but generally of the cow. It curdles spontaneously, as every one knows, sooner or later, according to the temperature of the atmosphere. When I mixed it with gastric juice, or any other fluid, I always left another portion by itself. In the former case, the coagulation soon took place without any sign of acidity, whereas milk alone required several hours, and sometimes a day or two, and the coagulum had always an acid taste.

CCXLIX. But it is time to consider the reasons adduced by others, to prove that digestion is attended with an incipient putrefaction. These reasons are founded upon facts related by different authors, and detailed in their order by Haller, in his great work (*a*). Nothing, according to them, can be more evident, than the signs of putrefaction during digestion. The stomach of a hyena and of a serpent, have been observed to emit an intolerable stench. The breath of the lion and eagle is very fœtid, as also that of the dog, when digestion has been prevented by the

(*a*) T. 6.

VOL. I.   Z   exhi-

exhibition of opium. A dog without taking opium, was obferved to emit an excrementitious odour from his ftomach; the food in the ftomach of birds has nearly that fmell. The fame obfervation has been applied to fifhes; and the inftance of a dog-fifh has been adduced, of which the ftomach was full of a fœtid jelly, that contained the food diffolved. The contents of the human ftomach fometimes become fœtid. Vegetable fubftances alfo degenerate into a putrid mafs, when they continue long in the ftomach, as appears from the putrid fmell they exhale, the green colour they impart to tincture of mallows, and the alcaline principle they afford on diftillation.

After having related thefe facts, the Swifs phyfiologift proceeds to give his own opinion. He thinks, that in digeftion there is only an incipient, not a complete putrefaction; which only takes place when the food remains a long time in the ftomach, as is evident from the facts juft mentioned. He alfo fuppofes, that the change produced by the digeftive powers, efpecially in the human ftomach, approaches nearer to putrefaction than acefcency; this he infers, from the putrid fmell that exhales from flefh found in the ftomach of fome animals, notwithftanding there has been

been no impediment to digeſtion (*a*). This opinion, adopted before Haller by Boerhaave (*b*), has moreover been received by two celebrated writers, Gardane (*c*), and Macquer (*d*).

CCL. Notwithſtanding the reſpectable authority of theſe authors, I do not think the facts adduced, ſufficient to perſuade an impartial philoſopher; they are not only too few, but were obſerved by mere accident; nor had the obſervers the ſmalleſt intention of entering into a full diſcuſſion of this point.

Though the time requiſite for digeſtion in different animals is different, yet in many it does not exceed five or ſix hours, and in ſome is ſtill ſhorter. Now it ſeemed proper to examine what change fleſh ſet to putrify, would undergo in that ſpace of time; I therefore took ſome freſh veal cut into ſmall pieces, and put it in a phial of water, which was ſtopped with paper. The phial was put into a ſtove, where the mercury roſe to between 30 and 35°.

About the beginning of the fourth hour, the fleſh had loſt its red colour, and was

(*a*) L. c.
(*b*) Chem. T. 2.
(*c*) Eſſai pour ſervir à l'Hiſtoire de la Putrefaction.
(*d*) Art. common Salt.

turning blue. It was alſo become flabby, but for nine hours it had no putrid ſmell. Mutton and beef, in ſeveral trials, did not anſwer to this time exactly, but no bad ſmell was ever perceptible for eight hours. Theſe experiments ſhew, that fleſh eaten by many animals, among which Man may be enumerated, has not time to run into the putrefactive fermentation, eſpecially as the temperature of animals is lower than that to which theſe ſeveral ſorts of meat were expoſed. However, for greater certainty, I made the following trials. I have before mentioned introducing into the ſtomachs of crows, pyriform glaſs veſſels, of which the ſmall end was open, and came out at the mouth (LXXXIX).

I now took two of them, and putting ſome beef, with a little water, into one, and ſome veal into the other, forced them down the throat of ſome crows. In order to examine the ſtate of the fleſh, I now and then drew them up, and immediately returned them. Between the ninth and tenth hours, the beef emitted an odour, which though it could not be called putrid, was diſagreeable. At the expiration of the tenth hour, there was a diſtinct putrid ſmell that became gradually ſtronger and ſtronger. In a day the fleſh turned livid, acquired a nauſeous taſte, and the

the particles began to feparate. The fame appearances took place rather fooner in the veal. It therefore appears, that flefh in the heat of this fpecies of bird requires a longer time to putrify, than to be digefted. After the glafs veffels were taken out of the ftomach, I gave one of them the fame quantity of beef and veal; and upon opening the ftomach in three hours, found that it was entirely confumed.

CCLI. Thefe experiments prove, that no putrid tendency is ever acquired by meat during digeftion. Nor did I ever perceive any fuch tendency in food lying in the ftomach (LXV, CCIX); yet as I had never made experiments for this exprefs purpofe, and as fome phyfiologifts adduce facts to prove the contrary (CCXLIX), I was under the neceffity of examining the ftomach of various animals, with this fole view.

Four hens were fed with kid, and in two hours one was killed: the ftomach was full; the flefh ftill retained its natural fweet favour, which at the furface was mixed with a bitterifh tafte, occafioned by its being impregnated with the gaftric fluid. It had no fmell, except that of this fluid. An hour afterwards the ftomach of another hen was examined; and here the flefh was beginning to be converted into

a gea gelatinous paſte, its ſmell was rather diſagreeable; I know not how to deſcribe it, but it was not at all penetrating, or putrid; the colour was ſtill reddiſh, it had not the leaſt nauſeous taſte, nor did it effervefce with acids, or change the colour of ſyrup of violets. Thus we ſee, it ſhewed no ſign either of incipient, or advanced putrefaction. In another hour the third hen was killed: the ſtomach contained a pultaceous maſs, more fluid than in the former cafe; but there was not the ſmalleſt token of putrefaction, any more than in the fourth hen, which was opened three hours afterwards, when the craw was empty, and the contents of the gizzard were now diſſolved.

CCLII. Some frogs juſt killed were ſet before two herons; the birds being hungry, devoured them greedily. In ſix hours one of them was opened; but whether the toughneſs of the ſkin retarded digeſtion, or whether that proceſs is ſlow in herons, the frogs had not loſt their ſhape; though the heads and limbs were either ſeparated, or on the point of being ſeparated from the trunk, and the fleſh was become very ſoft. The taſte, except the uſual bitterneſs, had nothing nauſeous, and the ſmell was by no means putrid. I waited five hours longer before I killed the
ſecond;

second; when I found but little flesh in the stomach, and that little was entirely decompofed, but did not emit the least putrid smell.

If fowls and herons afforded no token of putrefaction, much less could I expect it from the birds upon which my next trials were made; I mean, young owls, which digest flesh in three, or at most four hours. A young dog and cat were next fed, at the same time, with boiled beef. The former was opened in four hours and an half. The stomach was full of a mass of softened flesh, which emitted a very slight smell, exactly resembling the smell of the gastric fluid. The stomach of the cat was opened in five hours and an half, and was found to contain some remains of flesh, or rather a pulpy matter, which as in the former case, had the smell of gastric fluid. The flesh, when nearly digested, did not change the colour of syrup of violets, or effervesce with acids.

CCLIII. There are animals which retain the food in the stomach for a much longer time, as the falcon. Of that upon which I made so many experiments, I have already observed, that it would devour a whole pigeon at once, and continue without food the whole day afterwards (CLX). Whence, as also from its great bulk, we may infer, that the flesh remains a long time in the body before it is entirely digested. Some months after this was killed, another of a different species fell into my hands; it was of a larger size, and had no craw, so that the food passed immediately into the stomach. Notwithstanding it was pretty tame, and therefore valuable, yet I sacrificed it for the sake of these experiments, eighteen hours after it had devoured a chicken. What remained in the stomach weighed two ounces; it consisted of a pulp, in which the fibres could yet be discerned; but neither when subjected to the before-mentioned chemical trials, or when smelled and tasted, did it shew any sign of putrefaction. But among the animals that retain their food for a long time in the stomach, those with cold blood, and especially vipers, are, as we have seen, the most remarkable. A piece of lizard's tail preserved somewhat of its muscular structure, after having remained five days in the stomach of a land-snake (CXVIII). Three water snakes had not consumed all their food at the end of three days (CXXI). Another not even in six days (CXXV). A lizard remained sixteen days in the stomach of a viper, without losing its natural form (CXXVII). Other cold animals, such as eels, newts, and frogs,

must

muft not be forgotten. Four eels that had eaten fifh, retained a little after the expiration of three days, eighteen hours (CXXIX). On the fifth day, fome frogs had not quite digefted pieces of inteftine (CVI); which alfo happened to newts, two days after they had been fed with earth-worms (CVIII). But notwithftanding the food continued fo long in the ftomach of thefe feveral animals, I have exprefsly noticed, that it had not begun to putrify (CXXVII).

CCLIV. I have met only with two inftances which though they do not coincide with this invariable conftancy of nature, by no means detract from the certainty of the confequences that are to be deduced from it. Among the crows that were obliged to fwallow tubes for a confiderable length of time, fome fuffered in their healths, and became lean, though they were copioufly fupplied with food. But as they did not take it voluntarily, and as it was my wifh to keep them alive for the fake of experiments, I forced fome flefh down the throats of two, but to no purpofe; for they both died, one thirteen, the other fifteen hours afterwards. My curiofity led me to open them, and I found that the flefh continued whole and undigefted, and moreover that it was become putrid. But this evidently

arofe

arofe from the morbid condition of the animal, by which the gaftric fluid was altered, and rendered inefficacious. For this fpecies of birds, as I have feen in a hundred inftances, digefts flefh very fpeedily, and without any token of putrefaction appearing. It is alfo probable, that the putrid ftate of the food in the animals mentioned in the ccxlixth paragraph, arofe from their morbid condition; efpecially as it remained fo long in the ftomach of fome of them. It may alfo happen, that when an animal in health is killed and kept unopened a confiderable time, as often happens, the food in the ftomach may be found in a putrid ftate.

In the fame paragraph, the breath of the lion and eagle are faid to be fœtid. That of the lion I never had an opportunity of examining, but with the eagle it is far otherwife; for when I ftroaked the head gently, it would fometimes open its mouth, and raife a gentle cry; on thefe occafions it neceffarily made a long expiration, and in winter the breath appeared in the form of a little cloud. This cloud I have often fmelled, and caufed others to fmell, both when the bird was fafting, when the ftomach was full, and when the food was recently digefted; but it was never
fœtid,

fœtid, and indeed did not seem to have any kind of odour.

CCLV. The experiments described in the CCL, CCLI, CCLII, and CCLIIId paragraphs, not only shew that digestion is unaccompanied by putrefaction, but might induce us to suppose, that the stomach is provided with an antiseptic principle. Flesh inclosed in the pyriform glasses that were introduced into the stomach of crows, shewed evident signs of putrefaction in ten hours; whereas in eighteen they shew no appearance of the kind, when it is in immediate contact with the stomach (CCLIII). And although serpents, and the other amphibious animals above-mentioned (CCLIII) are of a cold temperature, yet in their temperature, which is nearly equal to that of the atmosphere, flesh becomes putrid in two days, sometimes in one, and sometimes even in a shorter time, while in their stomachs it remains untainted frequently for a much longer space. I could not therefore but conclude, that there is present in these cases, some cause that prevents the corruption which supervenes out of the body. What can this cause be? It was not difficult to detect it. I called to mind those unfinished digestions, which take place when flesh is immersed in gastric fluid contained in phials; where it is dissolved without ever

turning

turning putrid, notwithftanding it is kept long enough, and expofed to a fufficient heat. I could not then doubt, that the gaftric juices are at once the folvent, and the prefervative from putrefaction. Further reflection furnifhed me with proofs ftill more decifive. It appears from various paffages in the preceding differtations, that in attempts to produce artificial digeftion, little or no folution takes place, unlefs the fluid extracted from the ftomach is expofed to a confiderable heat (CXLII, CLXXXVI, CCI, CCXVII). But without this condition, it retains its antifeptic powers (CLXXXVI, CCXVII). Two phials, one containing fome gaftric fluid from a crow, and the other from a dog, together with fome veal and mutton, were kept thirty-feven days in winter, in an apartment without fire: the flefh was not either confumed or turned putrid; while fome that was immerfed in water, began to emit a fœtid fmell on the feventh, and about the thirtieth day was changed into a very offenfive liquamen. It is proper to add, that the gaftric fluid at laft lofes, though kept in phials ever fo clofely ftopped, its antifeptic quality, but it never becomes putrid itfelf. This at leaft I found to be the cafe with fome taken from a crow, and kept two months.

CCLVI. The

CCLVI. The discovery of this antiseptic property led me to enquire what would be the effect of immersing flesh, more or less putrid, in gastric fluid. Four portions that had an insupportable smell, were set in four bottles, which I filled with four different kinds of gastric fluid, viz. of a dog, a crow, an owl, and an eagle. This was done in March, and the bottles were kept twenty-five days in an apartment, where the heat was never less than 8, and never exceeded 12°. I could not perceive that it was at all more dissolved than if it had been immersed in water. With respect to the fœtid smell in the phials containing lamb and veal, it continued unchanged; but in the two others, which contained fowl and pigeon, it seemed rather diminished. This result suggested to me, that the gastric fluid might not only impede putrefaction, but restore putrified substances. I therefore repeated the experiment in June, and found that my suspicion was well founded. Some fowl and pigeon, in which putrefaction was pretty far advanced, were immersed in the gastric fluid of a dog and falcon, and remained in it thirty-seven hours, in which time they were reduced to a jelly, but had nearly lost their offensive smell. On comparing this with the preceding experiment,

I con-

I conjectured, that the superior efficacy of the gastric fluid in the latter case, proceeded from the warmth of the season: and this induced me to expose the same flesh under the same circumstances, to the sun about the middle of June. And now ten hours completely took away the fœtid smell. I did not neglect making the same experiment with the gastric fluid of other animals; the flesh generally lost its disagreeable odour, but sometimes for a reason, which I cannot assign, retained it in part. It is proper to add, that the recent fluid was always more efficacious than the old.

CCLVII. If we consider the CCLV and CCLVIth paragraphs, we must conclude, that putrid flesh loses this quality in the stomach of animals. Before I attempted to ascertain this point by experiment, nature herself gave me a decisive proof of it. At the time I kept a great number of fowls for my enquiries, I perceived that when they are allowed to eat at will, they cram their craw so full, that it is sixteen or twenty hours before it is completely evacuated. Curiosity led me to kill a cockrel that had about an ounce of meat, which happened to be bruised flesh, remaining in its craw: and I was struck with surprize, when I perceived that it had a strong putrid

putrid fmell; it was become foft, had a dull red colour, and a naufeous tafte: I immediately proceeded to examine the contents of the ftomach; but here I found the flefh quite decompofed with a bitter fweet tafte, and a fmell not in the leaft fœtid. The liquor therefore of the ftomach, had corrected the putrid quality which the flefh had acquired in the craw. The fame thing took place in fome hens. The flefh in the craw became putrid in fixteen hours, while that in the ftomach had no difagreeable odour. It fhould, however, be remarked, that the putrefactive fermentation never runs fo high within the craw, as it does out of the body, even when the heat is lefs ftrong. Whence I fufpected, that the fluid of the craw might alfo poffefs an antifeptic power, though in a degree very far inferior to that of the ftomach.

CCLVIII. I took a putrid piece of beef's lights, and dividing it into five portions, faftened a ftring to each, and then thruft them into the ftomachs of five ravens. The end of the ftring was brought out at the beak, as on former occafions (LXVIII), that I might be able to examine the flefh at pleafure. In three quarters of an hour, two of the pieces were drawn up: they were wafted, and at firft feemed to have loft their putrid fmell, but

upon

upon wiping off the gaftric fluid, it became again fenfible, but it was much diminifhed. Half an hour afterwards another piece, upon examination, was found to be ftill more wafted, and to have loft almoft all its bad odour, even when the gaftric fluid was carefully wiped away. In an hour afterwards the two remaining pieces were drawn up. They were reduced to the fize of a pea, and it would have been impoffible to tell that they had been ever putrid, fo perfectly were they recovered; even the tafte had nothing difagreeable, except the bitternefs which is always prefent on fuch occafions.

The great length of the neck prevented me from repeating this experiment upon the heron. I forced a femi-putrid frog, from which the fkin had been taken, into the ftomach, but I could not draw it up again; it was therefore neceffary to cut the ftring at the beak, when the bird immediately fwallowed it. It was my intention to kill the heron in about an hour, that I might examine what change the frog had undergone. But it was vomited before that time, probably on account of its being a difgufting food; for however greedily the heron devours living fifhes and frogs, it abftains from them when they are turning putrid. The gaftric
fluid

fluid had, notwithstanding, exerted both its antiseptic and solvent powers, during the forty-three minutes the frog had been in the stomach. Some tin tubes were then filled with putrid fish, and given to the animal; they were not, as before, thrown up, perhaps, because the putrid matter was not in contact with the stomach. The bird was killed three hours afterwards; what remained in the tubes weighed one-seventh of an ounce; it resembled a thick gelatinous paste, in which a few fleshy fibres might yet be distinguished, and which retained no vestige of its former putrid state.

CCLIX. I treated several small birds of prey, such as the two species of owl above described, and a young hawk, as I had done crows (CLVIII). They were fed with intestine, liver, and lungs of sheep, more or less putrid. Solution took place, and the putrefaction was corrected according to the time of the continuance of the flesh in the stomach. The hawk twice threw up what it had swallowed, probably, because its putrid state made it disagree with the stomach, for this never happened when it was fed with fresh meat. The gastric juices of the eagle produced the same effect upon flesh inclosed in tubes, and introduced into its stomach. Animals of cold blood

blood having very flow digeftive powers, were long in correcting the putrefaction of flefh. This effect, however, was at laft produced. The only precaution neceffary, was to return the fubftances into the ftomach when they were vomited, which often happened.

The laft experiments I made with this view, were upon a cat, a dog, and myfelf. I was obliged to force the putrid flefh down the throat of thefe animals, for notwithftanding they were exceedingly hungry, they obftinately refufed it. The dog retained what was forced upon him, but the cat vomited it along with a quantity of foam, and a liquor that appeared to be gaftric fluid. The flefh, which when it was given was exceedingly fœtid, had now loft its fmell entirely; of this, another cat, by eating it without afterwards throwing it up, gave me a clear proof. Upon opening the ftomach, I found the flefh half digefted, and with no other fmell than that which frefh meat ufually emits in like circumftances. In two hours and a half the dog was opened. The flefh lay in a little lake of gaftric fluid, nearly decompofed, nor did it either in tafte or fmell refemble tainted meat. The experiment I made on myfelf, confifted in fwallowing, at five different times, five tubes covered with linen, like thofe mentioned

tioned in the ccviiith paragraph: they were full of different forts of putrid flefh. I voided them feparately, and in each there was fome of the contents remaining, but not one exhibited the fmalleft token of putrefaction. Hence then it appears, that the various claffes of animals, and Man among the reft, in an healthy ftate, are endowed with the power, not only of checking the putrefaction of fubftances lodged in the ftomach, but alfo of correcting them when already putrid.

CCLX. By this difcovery I was led to reflect, that many animals living upon flefh, and matters that have a tendency to run into the putrefactive fermentation, never feed but upon fuch as are frefh and fweet; and that, if by any accident putrid food fhould get into the ftomach, they are fubject to vomiting and various bad fymptoms, and even death itfelf: fome inftances of vomiting excited by this cause, may be feen above (CCLVIII, CCLIX); while on the other hand, many animals delight in corrupted fubftances, as for inftance, the multitude of loathfome infects and worms that refide in fewers and fepulchres, and feed upon decaying carcafes. Among birds and quadrupeds, there are alfo fome that feek tainted flefh; fuch as the crow, the kite, the vulture, among the former; and among the

latter, the chacal and the hyena. While other animals fly the miafmata that arife from bodies in fuch a ftate, thefe feek and are guided by them to their abominable repafts. But now we are acquainted with the antifeptic virtue of the gaftric fluid, the difgufting manners of thefe animals ought no longer to furprize us, for the food, however putrid, muft be totally changed before it is converted into nutriment and animalized. And although the putrid quality is corrected by other animals, yet food in that ftate is noxious to them, on account of the difagreeable impreffion it makes on the organs of fmell and tafte, as alfo upon the ftomach, by which, and particularly by their noifome miafmata, the nervous fyftem is probably irritated. It befides feems likely, that the antifeptic power of the gaftric fluid of the former, is greater and more efficacious, and confequently that it more readily and more completely corrects putrefaction. Habit, which is juftly reputed a fecond nature, may bring animals, that naturally abominate putrid food, to live very well upon it. We have already feen the converfion of a pigeon from a granivorous into a carnivorous animal (CLXXV); and I brought it to eat not only frefh flefh, but fuch as was fœtid, and even completely putrified. The bird at firft abfolutely

folutely refufed it, and I was obliged to force it into the ftomach; for fome days it fuffered from this treatment, and became evidently leaner. But by degrees nature became inured to the food, and the pigeon, ftimulated by hunger, took it fpontaneoufly, till at laft it recovered its plumpnefs; and now its appetite for tainted, was as keen as it had been before for fweet meat. We may learn from this inftance, that cuftom is capable of changing difagreeable, and even noxious food, into good nourifhment.

CCLXI. But what fhall we fuppofe enables the gaftric fluid to check and correct putrefaction? As it contains a falt, and that of the ammonical kind (CCXLVI), and as befides the experiments of Pringle fhew, that all falts, whether acid, alkaline, or neutral, whether volatile or fixed, are antifeptic (*a*), it is obvious to conjecture, that thefe two qualities arife from the fame fource. I conceived, however, before I determined abfolutely, that it would be proper to attempt a few experiments. It is obferved by Pringle, that we muft employ common falt, which fo nearly refembles fal ammoniac, in confiderable quantity, if we wifh it to act as an antifeptic; other-

(*a*) Appendix, containing experiments on feptic and antifeptic fubftances.

wife, it is so far from checking, that it promotes putrefaction. Thus a drachm of salt, dissolved in two ounces of water, keeps meat sweet but a little while, and twenty-five grains a still less time; while ten, fifteen, and even twenty grains hasten its corruption. This paradox has been confirmed in France by the learned Mr. Gardane. Notwithstanding these authorities, I determined to bring the matter to the test of experiment. I therefore took four phials, and putting into each three pennyweights, six grains of fresh beef, pounded very small, I poured upon it an ounce and half of water. In the first phial were dissolved ten grains of common salt, in the second fifteen, in the third twenty, and the fourth was left without salt, as a term of comparison. The temperature of the place where they were kept, was about fifteen degrees. The first phial began first to emit a fœtid smell, the fourth next, then the second, and lastly the third. The other tokens of putrefaction appeared in the same order. When sal ammoniac was substituted in the place of common salt, the only difference in the result, consisted in the phial which contained no salt, and that which contained ten grains, beginning to exhale a putrid smell at the same time. It appears, therefore, that Pringle's experiment

periment was accurate, and that the fame thing nearly is true of fal ammoniac. In order to determine whether the antifeptic property of the gaftric fluid arifes from the fal ammoniac it contains, I diffolved a quantity of that falt by degrees in water, till it had nearly acquired the fame faltnefs as the gaftric fluid; fome bruifed flefh was then immerfed in it. That the water and the liquor of the ftomach had nearly the fame faltnefs, I affured myfelf, both by tafting it, and by dropping a few drops of each into a folution of filver in the nitrous acid, when each afforded the fame white precipitate. But it is this caufe that prevents putrefaction; for the flefh immerfed in the falt water, emitted a fœtid odour fooner than other flefh of the fame kind, infufed in common water: and although when more fal ammoniac was employed, putrefaction was retarded, it was not prevented; to attain this end, eighteen or twenty times as much falt as is contained in the gaftric fluid was requifite. Thefe facts feem clearly to fhew, that the antifeptic quality of the gaftric fluid does not depend on the fmall quantity of fal ammoniac it contains.

CCLXII. From the fceptic power of common falt in fmall quantity, Mr. Gardane deduces a confequence, which it may be proper to notice in paffing. He thinks the common falt we take with our food, being always in little dofes, forwards digeftion, by promoting putrefaction; upon which, according to him, as we have feen above (CCXLIX), that function depends. Though my numerous experiments completely deftroy this fuppofition, yet it feemed worth while to try what would happen to flefh feafoned with fuch a proportion of common falt as haftens putrefaction, and given to different animals. Some tubes, filled with flefh thus prepared, and others, with fome of the fame kind, without falt, were given to a dog and a cat. The animals were opened in five hours, and upon examining the tubes, I could not perceive that the falt had occafioned any difference. What remained undiffolved, had ftill a flight falt tafte, but not the leaft difagreeable fmell; and it was juft as much wafted as the other. It therefore appears, that this fmall dofe of falt had neither promoted digeftion, nor produced any tendency to putrefaction, being overpowered by the antifeptic quality of the gaftric fluid.

CCLXIII. But to return from this digreffion: If the falt contained in the gaftric fluid is not the caufe of its antifeptic power, to what other principle can it be owing? Macbride's bride's theory concerning the origin of this property in so many bodies, has great ingenuity. The cohesion and solidity of substances, is in his opinion owing to the fixed air they contain. Now when by any means this is taken away, the mutual adhesion of the several parts will be destroyed, and the body will either run into the putrefactive fermentation, or crumble into dust, according to the nature of its constituent parts. Hence it necessarily follows, that whatever substance has the power of impeding the separation of fixed air, or restoring it when separated, will also prevent or correct putrefaction. But antiseptic matters have, according to this physician, such a power. A piece of flesh, for instance, surrounded by a substance of this kind, is kept sweet, because the fixed air cannot make its escape; and that, probably, on account of its pores being blocked up by the finer particles of the antiseptic matter. Hence the flesh will long preserve its natural taste and consistence. If it has already become putrid, it will receive fixed air from the antiseptic body, and hence cease by degrees to exhale a fœtid smell, lose its fluidity and flabbiness, and at last recover its sweetness and firmness (a).

(a) Macbride, l. c.

Will not this theory account for the antiseptic power of the gaſtric fluid? Without going out of my way to examine the foundation on which it reſts, I will obſerve, that it ſeems by no means to afford the information wanted, ſince the gaſtric fluid is an antiſeptic of a ſingular ſort. Other ſubſtances poſſeſſing this property, while they keep away putrefaction, preſerve or reſtore the coheſion of the parts; whereas the gaſtric fluid being at once an antiſeptic and ſolvent, while it prevents or corrects putrefaction, reduces bodies into very ſmall particles. We muſt therefore conclude, that the property of this animal fluid ariſes from ſome other principle, though I cannot determine what that principle is, both for want of experimental data, and on account of the imperfect ſtate in which phyſicians have left the theory of putrefaction. I therefore choſe to acknowledge my ignorance, rather than invent ſome gratuitous hypotheſis; ſuch a mode of proceeding would ill agree with the diſpoſition of one, who has no other object in view than the diſcovery of truth.

CCLXIV. For the ſake of my readers, it may be proper to recapitulate what has been proved in this diſſertation. Firſt, of the three ſpecies of fermentation eſtabliſhed by modern chemiſts chemifts and naturalifts, viz. the fweet, the acetous, and the putrid, neither takes place in digeftion. Secondly, Though an acid sometimes appears during this procefs, yet it difappears entirely towards the conclufion of it. Thirdly, Putrefaction never in health attends digeftion. Fourthly, The gaftric fluid is a real antifeptic. I fuppofe my proofs, however conclufive, will not avail with thofe who eftablifh it as an axiom, that wherever there is heat and moifture, there muft be fermentation; and think that it muft therefore necessarily take place in the food, and not only in the ftomach and inteftines, but in the chyliferous and fanguiferous veffels: they indeed limit their doctrine fo far as to fay, that whereas out of the body it goes on rapidly, and with an inteftine commotion, in the body of it is flow, weak, and generally imperceptible. Let me intreat thefe learned and zealous advocates for fermentation to reflect, that my experiments are not directly repugnant to theirs. I only pretend to fhew, that not the fmalleft fenfible fermentation takes place in the ftomach of animals or Man. With refpect to fenfible fermentation, as it is amongft uncertain things, found logic forbids me alike either to admit or reject it.

APPEN-

APPENDIX.

ON THE

DIGESTION

OF THE

STOMACH AFTER DEATH.

By JOHN HUNTER, F. R. S. and Surgeon to ST. GEORGE's Hofpital*.

AN accurate knowledge of the appearances in animal bodies that die of a violent death, that is, in perfect health, or in a found ftate, ought to be confidered as a neceffary foundation for judging of the ftate of the body in thofe that are difeafed.

But as an animal body undergoes changes after death, or when dead, it has never been fufficiently confidered what thofe changes are; and till this be done, it is impoffible we

* See Philofophical Tranfactions, Vol. LXII. p. 447.

should judge accurately of the appearances in dead bodies. The diseases which the living body undergoes (mortification excepted) are always connected with the living principle, and are not in the least similar to what may be called diseases or changes in the dead body: without this knowledge, our judgment of the appearances in dead bodies must often be very imperfect, or very erroneous; we may see appearances which are natural, and may suppose them to have arisen from disease; we may see diseased parts, and suppose them in a natural state; and we may suppose a circumstance to have existed before death, which was really a consequence of it; or we may imagine it to be a natural change after death, when it was truly a disease of the living body. It is easy to see therefore, how a man in this state of ignorance must blunder, when he comes to connect the appearances in a dead body with the symptoms that were observed in life; and indeed, all the usefulness of opening dead bodies depends upon the judgment and sagacity with which this sort of comparison is made.

There is a case of a mixed nature, which cannot be reckoned a process of the living body, nor of the dead; it participates of both,

both, inafmuch as its caufe arifes from the living, yet cannot take effect till after death.

This fhall be the object of the prefent paper; and, to render the fubject more intelligible, it will be neceffary to give fome general ideas concerning the caufe and effects.

An animal fubftance when joined with the living principle, cannot undergo any change in its properties but as an animal; this principle always acting and preferving the fubftance, which it inhabits, from diffolution, and from being changed according to the natural changes, which other fubftances, applied to it, undergo.

There are a great many powers in nature, which the living principle does not enable the animal matter, with which is is combined, to refift, viz. the mechanical and moft of the ftronger chemical folvents. It renders it however capable of refifting the powers of fermentation, digeftion, and perhaps feveral others, which are well known to act on this fame matter, when deprived of the living principle, and entirely to decompofe it. The number of powers, which thus act differently on the living and dead animal fubftance, is not afcertained: we fhall take notice of two, which can only affect this fubftance when deprived of the living principle; which are,

putre-

putrefaction and digeftion. Putrefaction is an effect which arifes fpontaneoufly; digeftion is an effect of another principle acting upon it, and fhall here be confidered a little more particularly;

Animals, or parts of animals, poffeffed of the living principle, when taken into the ftomach, are not the leaft affected by the powers of that vifcus, fo long as the animal principle remains; hence it is that we find animals of various kinds living in the ftomach, or even hatched and bred there: but the moment that any of thofe lofe the living principle, they become fubject to the digeftive powers of the ftomach. If it were poffible for a man's hand, for example, to be introduced into the ftomach of a living animal, and kept there for fome confiderable time, it would be found, that the diffolvent powers of the ftomach could have no effect upon it; but if the fame hand were feparated from the body, and introduced into the fame ftomach, we fhould then find that the ftomach would immediately act upon it.

Indeed, if this were not the cafe, we fhould find that the ftomach itfelf ought to have been made of indigeftible materials; for, if the living principle was not capable of preferving animal fubftances from undergoing

ing that procefs, the ftomach itfelf would be digefted.

But we find on the contrary, that the ftomach, which at one inftant, that is, while poffeffed of the living principle, was capable of refifting the digeftive powers which it contained, the next moment, viz. when deprived of the living principle, is itfelf capable of being digefted, either by the digeftive powers of other ftomachs, or by the remains of that power which it had of digefting other things.

From thefe obfervations, we are led to account for an appearance which we often find in the ftomachs of dead bodies; and at the fame time they throw a confiderable light upon the nature of digeftion. The appearance which has been hinted at, is a diffolution of the ftomach at its greateft extremity; in confequence of which, there is frequently a confiderable aperture made in that vifcus. The edges of this opening appear to be half diffolved, very much like that kind of diffolution which flefhy parts undergo when half digefted in a living ftomach, or when diffolved by a cauftic alkali, viz. pulpy, tender, and ragged.

In thefe cafes, the contents of the ftomach are generally found loofe in the cavity of the abdomen, about the fpleen and diaphragm.

In many fubjects this digeftive power extends much further than through the ftomach. I have often found, that after it had diffolved the ftomach at the ufual place, the contents of the ftomach had come into contact with the fpleen and diaphragm, had partly diffolved the adjacent fide of the fpleen, and had diffolved the diaphragm quite through; fo that the contents of the ftomach were found in the cavity of the thorax, and had even affected the lungs in a fmall degree.

There are very few dead bodies, in which the ftomach is not, at its great end, in fome degree digefted; and one who is acquainted with diffections, can eafily trace the gradations from the fmalleft to the greateft.

To be fenfible of this effect, nothing more is neceffary, than to compare the inner furface of the great end of the ftomach, with any other part of the inner furface; what is found, will appear foft, fpongy, and granulated, and without diftinct blood-veffels, opake and thick; while the other will appear fmooth, thin, and more tranfparent; and the veffels will be feen ramifying in its fubftance, and upon fqueezing the blood which they contain from the larger branches to the fmaller, it will be found to pafs out at the digefted ends

ends of the veffels, and appear like drops on the inner furface.

Thefe appearances I had often feen, and I do fuppofe that they had been feen by others; but I was at a lofs to account for them; at firft, I fuppofed them to have been produced during life, and was therefore difpofed to look upon them as the caufe of death; but I never found that they had any connection with the fymptoms: and I was ftill more at a lofs to account for thefe appearances, when I found that they were moft frequent in thofe who died of violent deaths, which made me fufpect that the true caufe was not even imagined (*a*).

At this time I was making many experiments upon digeftion, on different animals, all of which were killed, at different times, after being fed with different kinds of food;

fome

(*a*) The firft time that I had occafion to obferve this appearance in fuch as died of violence and fuddenly, and in whom therefore I could not eafily fuppofe it to be the effect of difeafe in the living body, was in a man who had his fkull fractured, and was killed outright by one blow of a poker. Juft before this accident, he had been in perfect health, and had taken a hearty fupper of cold meat, cheefe, bread, and ale. Upon opening the abdomen, I found that the ftomach, though it ftill contained a good deal, was diffolved at its great end, and a confiderable part of thefe its contents lay loofe in the general cavity of the belly. This appearance puzzled me very some of them were not opened immediately after death, and in some of them I found the appearances above described in the stomach. For, pursuing the enquiry about digestion, I got the stomachs of a vast variety of fish, which all die of violent deaths, and all may be said to die in perfect health, and with their stomach commonly full; in these animals we see the progress of digestion most distinctly; for as they swallowed their food whole, that is, without mastication, and swallow fish that are much larger than the digesting part of the stomach can contain (the shape of the fish swallowed being very favourable for this enquiry), we find in many instances that the part of the swallowed fish which is lodged in the digesting part of the stomach is more or less dissolved, while that part which remains in the œsophagus is perfectly sound.

And in many of these I found, that this digesting part of the stomach was itself re-

---

very much. The second time was at St. George's Hospital, in a man who died a few hours after receiving a blow on his head, which fractured his skull likewise. From those two cases, among other conjectures about so strange an appearance, I began to suspect that it might be peculiar to cases of fractured skulls; and therefore, whenever I had an opportunity, I examined the stomach of every person who died of that accident: but I found many of them which had not this appearance. Afterwards I met with it in a soldier who had been hanged.

duced

duced to the fame diffolved ftate as the digefted part of the food.

Being employed upon this fubject, and therefore enabled to account more readily for appearances which had any connection with it, and obferving that the half-diffolved parts of the ftomach, &c. were fimilar to the half-digefted food, it immediately ftruck me, that it was from the procefs of digeftion going on after death, that the ftomach, being dead, was no longer capable of refifting the powers of that menftruum, which itfelf had formed for the digeftion of its contents; with this idea, I fet about making experiments to produce thefe appearances at pleafure, which would have taught us how long the animal ought to live after feeding, and how long it fhould remain after death before it is opened; and above all, to find out the method of producing the greateft digeftive power in the living ftomach: but this purfuit led me into an unbounded field.

Thefe appearances throw confiderable light on the principles of digeftion; they fhew that it is not mechanical power, nor contractions of the ftomach, nor heat, but fomething fecreted in the coats of the ftomach, which is thrown into its cavity, and there

there animalifes the food (*a*), or affimilates it to the nature of the blood. The power of this juice is confined or limited to certain fubftances, efpecially of the vegetable and animal kingdoms; and although this menftruum is capable of acting independently of the ftomach, yet it is obliged to that vifcus for its continuance.

(*a*) In all the animals, whether carnivorous or not, upon which I made obfervations or experiments to difcover whether or not there was an acid in the ftomach, (and I tried this in a great variety), I conftantly found that there was an acid, but not a ftrong one, in the juices contained in that vifcus in a natural ftate.

EXPE-

EXPERIMENTS

CONCERNING

DIGESTION.

TRANSLATED FROM THE

INAUGURAL DISSERTATION

OF

Dr. STEVENS.

Publiſhed at Edinburgh in 1777.

THE following experiments were made at Edinburgh upon an Huſſar, a man of weak underſtanding, who gained a miſerable livelihood, by ſwallowing ſtones for the amuſement of the common people, at the imminent hazard of his life. He began this practice at the age of ſeven, and has now followed it twenty years. His ſtomach is ſo much diſtended, that he can ſwallow ſeveral ſtones at a time; and theſe may not only be plainly felt, but may be heard, whenever the hypogaſtric region is ſtruck.

EXPERIMENT I.

At eight o'clock in the evening, I gave the subject of my experiments a hollow silver sphere, divided into two cavities by a partition, and perforated on the surface with a great number of holes, capable of admitting a needle: into one of these cavities was put four scruples and a half of raw beef, and into the other five scruples of raw bleak. The sphere was voided in twenty-one hours, when the beef was found to have lost one scruple and a half, and the fish two scruples. The rest was much softened, but had no disagreeable smell.

II. A few days afterwards he took the same sphere, containing in one cavity a scruple and four grains of raw beef, and in the other four scruples and eight grains of the same boiled. In forty-three hours the sphere was returned, and the raw flesh had lost one scruple and two grains, and the boiled one scruple and sixteen grains.

III. Suspecting that if these substances were divided, so that the solvent could have freer access to them, more of them would be dissolved. I procured another sphere with holes, so large as to receive a crow's quill, and enclosed some beef a little masticated in it.

it. It was voided quite empty, thirty-eight hours after it was fwallowed.

IV. Seeing how readily the chewed meat was diffolved, I thought of trying whether it would be as foon diffolved in a fphere with large holes, but without being chewed. I therefore put a fcruple and eight grains of pork into one cavity, and into the other the fame quantity of cheefe. The fphere was retained forty-three hours, at the end of which not the fmalleft remains of either pork or cheefe could be found.

V. He afterwards fwallowed the fame fphere, containing in one partition fome roafted turkey, and in the other fome boiled falt herring. In forty-fix hours it was voided, and nothing of the turkey or herring now appeared, both having been completely diffolved.

VI. Having found that animal fubftances, though inclofed in tubes, are eafily concocted, I next determined to try whether vegetables, which are more difficulty digefted, would be fo too. I therefore enclofed an equal quantity of raw parfnep and potatoe in a fphere. It was voided after having continued forty-eight hours in the alimentary canal, when both fpecies of vegetable were found to be diffolved.

VII. Pieces

VII. Pieces of apple and turnep, both raw and boiled, were diſſolved in thirty-ſix hours.

VIII. He next ſwallowed ſome grains of wheat, rye, barley, oats, and peaſe, contained in a ſphere, which remained ſeveral hours in the alimentary canal, but no alteration was produced on any of its contents, except upon the peaſe, which were ſwoln, and burſt by the humidity they had imbibed.

IX. The readineſs with which the gaſtric fluid had acted upon roaſted animal ſubſtances, induced me to try what change would be produced by it upon hard ones, ſuch as bone. I therefore incloſed in one partition of a ſphere, ſome of the bone from a leg of mutton, and in the other part of a turkey's wing. The ſphere was retained forty-eight hours. The bone was weighed, and found to have loſt nothing of its weight, while the fleſh, ſkin, and ligaments were quite diſſolved, ſo that the bones of the wing were now quite ſeparate; but they had undergone no perceptible alteration.

X. Inanimate matters being ſo readily ſoluble, I reſolved to enquire how far living animals are capable of reſiſting the action of this powerful menſtruum. With this view, an animal ſuppoſed to be deſtitute of pores, and,

and, according to my experiments, capable of fuftaining a degree of heat equal to the human temperature, was enclofed in a fphere perforated with fmall holes, to prevent the leech from wounding the ftomach. The Huffar took it, and voided it about the ufual time, when nothing was found except a black vifcid miafma, the remains of the digefted leech. This experiment was repeated with earth-worms, and they were diffolved with equal facility. But as they cannot fo well fupport the human temperature, it is probable they died before they began to be diffolved *.

It was my intention to make more experiments of this kind, but as the Huffar left Edinburgh foon afterwards, I was obliged to have recourfe to dogs and ruminating animals.

XI. A whelp, three months old, having been kept fourteen hours without food, was forced to fwallow four oval ivory globes, of different fizes, and perforated with many fmall holes. One contained beef, another haddock, a third potatoe, and a fourth cabbage, all raw, and weighing each fixteen grains. In four hours the animal was killed and opened. The globes were found in the

* Perhaps this is alfo the cafe with leeches.

ftomach, and their refpective contents were diminifhed in the following proportions: The fifh had loft nine grains, the beef five, the potatoe three, and the cabbage one. The globes themfelves appeared to be thinner, but as I had no fufpicion that the ivory would be affected by the gaftric fluid, I did not weigh them before the experiment. I could not therefore exactly afcertain their diminution.

XII. Having procured a whelp five months old, it was kept fafting fixteen hours, and then four of the globes ufed in the foregoing experiment, each containing a certain quantity of mutton, turbot, parfnep, and potatoe were forced upon it. Thefe fubftances had been previoufly expofed to the action of fire, and each weighed fixteen grains. Seven hours afterwards the animal was killed, and the globes were taken out of the ftomach; when the fifh was found to have loft ten grains and a half, the mutton fix, the potatoe five, and the parfnep nothing. The fpheres were become ftill thinner, but I had as before, neglected to weigh them.

XIII. A dog fix months old was kept fafting the ufual time, and the fame four fpheres were given him. The firft contained fixteen grains of boiled mutton, the fecond as much boiled fifh, the third the fame quantity of

boiled

boiled potatoe, and the fourth of boiled parſnep. In eight hours it was killed and opened.
The globes were found greatly altered. The
extreme parts, not the middle, were totally
diſſolved, ſo that the contents lay looſe in
the ſtomach. The ſpheres, before the experiment, weighed together three ſcruples
ſixteen grains; the fragments weighed only
one ſcruple and twenty grains. The mutton
and fiſh were entirely concocted, the potatoe
had loſt twenty-one grains; but the parſnep
was unchanged.

XIV. Being ſurprized at the ſpeedy ſolution of ivory by the gaſtric fluid, I determined to ſubject other hard bodies to its
action. I therefore carefully weighed three
pieces of a ſheep's thigh bone, and gave
them to a dog that had been long kept faſting. Seven hours afterwards the animal was
killed, and the bones were taken out of the
ſtomach. The firſt had loſt ſeven, the ſecond nine, and the third twelve grains. The
ſolution began at the internal ſurface, and
advanced towards the center, ſo that the cavity was conſiderably augmented (*a*).

I more-

(*a*) In order to aſſure myſelf that this ſolution was not owing
to fermentation, or an acid, I immerſed a bone of the ſame
kind, in an alimentary mixture, conſiſting of roaſted beef,
wheaten

I moreover obliged my dog to swallow pieces of cartilage, but I found that the gastric fluid produced no effect upon them.

xv. As the ivory spheres and bones were so readily dissolved in the foregoing experiments, I was induced to make trial of some bodies still harder. With this view I procured some cylindrical tin tubes, perforated with a great number of holes; of which four were given to a dog that had been kept fasting twelve hours. The first contained sixteen grains of roasted beef, the second the same quantity of veal, the third of fat, and the fourth of wheaten bread. In ten hours the animal was killed and opened, and the tubes were taken out of its stomach. The beef and bread were quite dissolved; the veal had lost only ten grains, and the fat eight and a half. The tubes had not undergone the smallest alteration.

xvi. As in the last experiment the veal was not so soon dissolved as the beef, I began to suspect that the flesh of young animals in general is less easy to digest than that of old

---

wheaten bread and water, beaten into a pulp. When it had remained forty-eight hours in a temperature, equal to 102 deg. of Fahr. Therm. it was examined: the fermentation had run very high, and the acidity was strong, but the bone had undergone no diminution. It was, however, much softened.

ones.

ones. I therefore took care to repeat the experiment with lamb and mutton, which were put in equal quantities into two tubes. The refult was as before. In feven hours the mutton was quite diffolved, whereas the lamb had loft only ten grains. The remains of veal and lamb in thefe experiments were furrounded with a vifcid gelatinous matter.

XVII. Sixteen grains of raw beef, and the fame quantity of roafted were inclofed in two tubes, and given to a dog, which was killed feven hours afterwards, when the former was found to have loft fifteen grains, while the latter was completely diffolved.

XVIII. The fame experiment was repeated with fifh inftead of flefh. Sixteen grains of raw and as much boiled haddock, were enclofed in two tubes, and given to a dog. When he was killed, no remains of the boiled could be found; the raw portion had loft fourteen grains.

XIX. I next enquired whether quadrupeds or birds are moft eafily digefted. For this purpofe, equal quantities of beef, mutton, and fowl were inclofed in three tubes, and given to a dog; they were each roafted, and weighed fixteen grains. Upon killing the dog, and examining the tubes, I found that the

the mutton and beef had been diffolved, while the fowl had only loft eleven grains.

Moft of the experiments related above, were repeated oftner than once, and afforded the fame refult. We cannot therefore entertain any doubt concerning the mode of digeftion in this clafs of animals. Whether the concoction of ruminating animals is effected in the fame manner, I endeavoured to afcertain by the following experiments.

xx. I gave a fheep four cylindrical tin tubes, each containing fixteen grains of raw beef, falmon, turnep, or potatoe; fix hours afterwards the animal was killed; the tubes were found in the firft ftomach. The fifh and flefh were unaltered, whereas the turnep and potatoe were quite diffolved.

xxi. The fame experiment being repeated with the fame fubftances boiled, afforded the fame refult. The vegetables were digefted, and the beef and falmon unchanged.

xxii. Having found that the fheep digefts vegetables very readily, but is incapable of diffolving animal fubftances, I had next recourfe to the ox. Four tubes, one containing raw beef, another fifh, a third chopped hay, and the fourth leaves of pot-herbs, were given to an animal of this fpecies, and it was killed ten hours afterwards. The tubes

lay

lay in the firſt ſtomach; the fiſh and fleſh were not altered; but I could find no remains of the hay or herbs.

Many experiments of the ſame kind were made upon this animal, and they led me to the ſame concluſion, viz. that the gaſtric fluid of the ox kind, eaſily and ſpeedily diſ-ſolves vegetables, but is incapable of producing this effect upon animal ſubſtances.

In all theſe experiments, I attribute the ſolution of the food to a powerful menſtruum ſecreted by the coats of the ſtomach. It may be objected, that my experiments do not clearly ſhew whether the food is concocted by the gaſtric fluid, or by fermentation, for both cauſes may act equally upon aliment incloſed in the ſpheres. But beſides the arguments already adduced to ſhew (*a*), that fermentation does not produce this effect, many circumſtances attending theſe experiments clearly ſhew the efficacy of the gaſtric liquor. For in the experiments in which the food was not quite diſſolved, the ſolution always began at the ſurface, and proceeded towards the center, and what remained, ſhewed no tokens of fermentation.

(*a*) In the part that has been omitted.

In the xiiith experiment ivory was diffolved, while parfnep, a vegetable of foft texture, and liable to fermentation, was not at all altered.

To remove every doubt, the following experiment was feveral times repeated, and always afforded the fame refult.

xxiii. Having kept a dog fafting eighteen hours, that his ftomach might be free from the remains of food, I killed it, and collected about half an ounce of pure gaftric fluid, which was put into a phial with twelve grains of roaft beef. The fame quantity of the fame beef was put into another phial, containing water, in order to ferve for a term of comparifon. Both phials were placed in a furnace, of which the temperature was equal to 102--104° of Fahrenheit's thermometer. In eight hours the beef in the gaftric fluid was quite diffolved, whereas that in the water had undergone no perceptible alteration. In twenty-four hours both phials were taken out of the furnace and carefully examined. The food diffolved in the gaftric fluid emitted a rancid and pungent, but by no means a putrid odour; it refembled very much the fmell of burnt feathers. The meat in the other phial was quite putrid, and intolerably

tolerably fetid; but its bulk was not diminifhed.

I carefully obferved the phial containing the gaftric fluid during the folution, but could perceive no air-bubbles arifing, or any other token of fermentation. I repeated this experiment with mafticated meat, when the folution was much more fpeedily completed.

I afterwards made trial of mutton, veal, lamb, and other animal, together with a great variety of vegetable fubftances; all were eafily diffolved; but the time requifite for the completion of this procefs was different, and anfwered exactly to the refults of the preceding experiments.

As in this experiment there was no fign of fermentation or putrefaction, I fufpected that the gaftric fluid, as well as the faliva (a), retards both the one and the other. In order to determine this, I made the following experiment.

XXIV. I took two alimentary mixtures, each confifting of mutton and bread in equal quantities. Upon one, half an ounce of the recent gaftric juice of a dog was poured, and

(a) Where did the author learn that the faliva checks fermentation? It appears from all the experiments that have been made to forward it. T.

upon the other the fame quantity of pure water. Both mixtures were beaten to a pulp, and inclofed in phials accurately ftopped; they were then fet in a furnace, heated to the 102d deg. of Fahrenheit's thermometer. Fermentation took place in a few hours in the phials that contained the water, the folid contents rofe to the furface, and air was extricated with a confiderable inteftine motion. The mixture immerfed in the gaftric fluid, remained fourteen hours with fcarce any tokens of fermentation; but a fhort time afterwards, this procefs evidently took place. The bread and flefh arofe to the furface of the mixture, a fediment began to be depofited, and air-bubbles were continually extricated. But thefe phænomena continued much longer than in the other phial; the commotion was lefs violent, and the air was not fo rapidly extricated. When the fermentation had entirely ceafed, the tafte of the mixture in this phial was indeed acid, but not fo ftrong as in the other, and it was converted into a fluid by the folvent power of the gaftric liquor.

xxv. I divided a piece of putrid mutton into two parts, each of which was put into a feparate phial, and to one, half an ounce of the recent gaftric fluid of a dog was added,

and

and to the other, which was defigned as a term of comparifon, as much water. They were fet in a cool place, and two days afterwards I examined them, when the latter emitted an intolerably putrid fmell, and the other, though it had yet a bad odour, did not fmell fo difagreeably as the preceding, nor even fo difagreeably as at firft. Upon fhaking the phial, the meat fell to pieces, but it was not quite diffolved. This, perhaps, happened, becaufe it was not expofed to a fufficient heat.

Thefe experiments throw great light on digeftion. They fhew, that it is not the effect of heat, trituration, putrefaction, or fermentation alone, but of a powerful folvent, fecreted by the coats of the ftomach, which converts the aliment into a fluid, refembling the blood. If it fhould be afked, what defends the organ itfelf, I would anfwer, that it is the vital principle, as Mr. Hunter's (a) obfervations fhew; after death it is diffolved as readily as any other inanimate fubftance. It is probable, that every fpecies of

---

(a) Philofoph. Tranf. for 1772. The ingenious obferver feems, however, to attribute too much to this principle. He fuppofes, that whatever poffeffes it, is capable of refifting the action of the gaftric liquor; his arguments by no means prove this.

of animal has its peculiar gaſtric liquor, capable of diſſolving certain ſubſtances only. Some living ſolely upon vegetables, others upon animals, and theſe cannot be obliged to feed upon plants, by a faſt of whatever continuance. All, by an infallible inſtinct, chooſe what is beſt adapted to their gaſtric fluid. The food, when diſſolved, is expelled from the ſtomach, and being mixed with the bile and pancreatic juice in the duodenum, is changed into a mild blood and in-

this. Worms, indeed, live in the human ſtomach, but it does not follow, that other animals alſo can, for nature may have given them a particular ſtructure of body. The following conſiderations will render the general propoſition very doubtful. Fiſhes ſwallow and digeſt living crabs, lobſters, &c. The leech is concocted by the human ſtomach, though it has no pores, and can ſuſtain a temperature equal to that of man. Cornelius found a ſnake half digeſted in a bird's ſtomach, but ſtill alive. Plot ſaw one eye conſumed, while the fiſh was alive. It ſeems therefore probable, that the gaſtric liquor acts alſo upon living things. Perhaps, likewiſe, it is ſometimes ſo changed, as to act on the ſtomach itſelf. The following caſes communicated by Dr. Monro render this probable. A lady, that uſed to complain of pain in the ſtomach, died ſuddenly. Upon opening the body, a hole was found in the left ſide, and the coats were relaxed as if they were half putrified. There were no appearances of gangrene. A boy died after having long ſtruggled with ſimilar pains. The ſtomach exhibited the very ſame appearance, if we except the hole. From the preceding ſymptoms, one may venture to ſuppoſe, that ſome alteration was produced before death. But this is only conjecture, and future experiments muſt determine the queſtion.

odorous

odorous liquid, which is denominated chyle. The chyle is abforbed by numberlefs veffels, and is carried by the thoracic duct into the fubclavian vein, in order to repair the conftant wafte of the body,

AN
ANALYTICAL INDEX
OF THE
CONTENTS
OF THE SIX PRECEDING
DISSERTATIONS.

INTRODUCTION.

REASONS of the author for treating concerning Digestion. Syftems relative to this function   Page i

DISSERTATION I.

ON THE DIGESTION OF ANIMALS WITH MUSCULAR STOMACHS, COMMON FOWLS, TURKEYS, DUCKS, GEESE, DOVES, PIGEONS.

I. Owing to mere trituration, according to feveral authors. This opinion extended by them to all other animals   3
II. Reaumur's experiments on one kind of grain, whence he infers, that the reduction of the food to pieces is the effect of trituration alone in birds with mufcular ftomachs or gizzards   —   4
III. Reaumur's experiments extended by the author to other grain   — —   5
IV. Variation of thefe experiments   —   6
V. Other variations of them by previoufly macerating the grain in the craw of gallinaceous fowls   ib.
VI. By taking off the fkin   —   7
VII. The fame fubjected to frefh trials in ducks, turkies, geefe, doves, and pigeons   —   ib.
VIII. Conclufion   — —   8

IX.

INDEX.

IX. A neceffary precaution — p. 8
X. Tin tubes broken and diftorted in the gizzards of turkies - 9
XI. ——liable to the fame accidents, though ftrengthened with iron wire, injuries ftill more furprizing 10
XII. A full confirmation of the Florentine experiments, concerning the trituration of empty glafs globules in the ftomach of gallinaceous fowls, the longer they remain in the ftomach, the more finely are they pulverized. The facility with which they are broken, in proportion to the fize of the animal - 11
XIII. Profeffor Pozzi denies the trituration of thefe balls - - - 12
XIV. Vallifneri miftaken, in fuppofing that thefe effects are produced by the gaftric fluid - 13
XV. Pieces of glafs lofe their edges and points by continuing in the gizzard of a cock - 14
XVI. Angles of a large garnet abraded in the gizzard of a pigeon - - - 15
XVII. Coats of the gizzard not hurt by thefe fubftances - - - 16
XVIII. Large needles fixed in a leaden ball, broken by the action of the gizzard of a turkey, without injury to that organ. Injury done to the ball - ib.
XIX. Lancets broken in the fame manner - 17
XX. Time requifite for thefe effects to be produced 18
XXI. The gizzard of young fowls fometimes a little hurt by the metallic points - - ib.
XXII. Why does not the gizzard fuffer from thofe fubftances - - - 19
XXIII. Whether, as fome fuppofe, becaufe the pebbles that are always found in it reduce thefe hard fubftances to pieces - - - 20
XXIV. This is a mere hypothefis, and fhould be verified by experiment - - ib.
XXV, XXVI. No foundation for the opinion of the Florentine academicians, that hard bodies are more eafily broken, the more pebbles the bird has in its gizzard. - - - 21, 22
XXVII. Means contrived for afcertaining the ufe of thefe pebbles - - - 22

XXVIII.

XXVIII. When most of them are come away, the effects of trituration not at all diminished. The gizzard not hurt by sharp bodies when it contains no stones 23
XXIX. Pebbles in the gizzard of nestlings - 25
XXX. When it is proper to examine these birds, in order to find their gizzards without stones. Those without them break down hard and short bodies without sustaining any injury - - ib.
XXXI. The question decided, whether digestion depends on these stones - - 26
XXXII. Decision of other curious questions 27
XXXIII. Fowls of this class do not seem to seek them from design, but only to swallow them because they are mixed among their food - 28
XXXIV. Trituration is the immediate effect of the gastric muscles - - 30
XXXV. Nature of the internal coat of the gizzard. It is divided by drawing sharp substances over it ib.
XXXVI. Not so when they are inclosed and agitated by the hand - - - 31
XXXVII. Reaumur's observations on the living gizzard. Slight motion of it - - 32
XXXVIII. Similar motion observed by the author 33
XXXIX. Whether the gastric muscles also change the food into that pultaceous mass called *chyme*. Facts affording room to suspect, that the gastric fluid produces this effect - - 34
XL. Other facts that add strength to these suspicions 35
XLI. Decisive experiments in favour of this opinion 36
XLII, XLIII. Others equally decisive--Precaution 38--40
XLIV, XLV. How an experiment of Reaumur ought to be understood - - 40--42
XLVI. To understand digestion thoroughly, it is necessary to examine the œsophagus and gizzard likewise. Description of the œsophagus of a goose - 44
XLVII. Numerous follicles of different sizes in it. Excretory ducts, and the fluid that oozes out - 45
XLVIII. Description of the stomach. Largeness of the muscles. Their action. Cartilaginous coat 47
XLIX. Œsophagus and gizzard of the fowl and turkey. Follicles. Fluid. Craw, and its glands ib.

L. Œso-

L. Œfophagus and gizzard of other gallinaceous fowls - - 48
LI. No appearance of glands in the gizzard. Whether any fluid can come into it by any other means. Suspicion of Reaumur on this fubject. Experiments 49
LII. Liquor falls in plenty out of the œfophagus into the ftomach - - - 51
LIII. Bitternefs of the gaftric fluid occafioned by the bile - - - 52
LIV. Maceration in the gizzard, the firft ftep towards digeftion. The manner in which it paffes from the craw to the gizzard - - 53
LV. No trituration in the craw. Changes the food undergoes in the gizzard - 54
LVI. Artificial digeftion. The gaftric fluid more efficacious than water - - 55
LVII. The fame. Neceffary precaution - 58

DISSERTATION II.

ON THE DIGESTION OF ANIMALS WITH INTERMEDIATE STOMACHS, CROWS, HERONS.

LVIII. In what fenfe crows can be called animals with intermediate ftomachs - 59
LIX. Ufe of experiments on crows, becaufe they, like man, are omnivorous. Very convenient on account of their throwing up indigeftible bodies - 61
LX. The ftones in the ftomach more eafily evacuated from crows than gallinaceous birds. Not requifite for digeftion. Swallowed only becaufe they happen to be mixed among the food. - 62
LXI. Gaftric fluid incapable of diffolving entire grains 63
LXII. But it diffolves bruifed ones. The mechanical action of the ftomach does not contribute to this 64
LXIII. Variation of thefe experiments - 65
LXIV. Tender vegetables eafily and foon diffolved by crows - - - 66
LXV. Flefh diffolved without the concurrence of mufcular action. Manner of action of the gaftric fluid 67

LXVI,

INDEX.

LXVI, LXVII, LXVIII. Experiments fhewing, that the digeftion of the flefh is nearly proportional to the quantity of gaftric fluid by which it is invefted   69, 70--72
LXIX. Gaftric fluid of neftlings more efficacious than that of adult ones   -   -   74
LXX, LXXI, LXXII, LXXIII. An error of Cheyne. Gaftric fluid of crows incapable of diffolving hard bones.-- Diffolves tender ones   -   -   75--79
LXXIV. Whether the œfophagus of this bird will diffolve flefh like that of fome fifhes   -   79
LXXV. Œfophagus of the crow defcribed. The follicles and the fluid   -   -   80
LXXVI. The ftomach defcribed---its glands and their liquor   -   -   81
LXXVII. Œfophageal juice produces fome concoction   -   -   82
LXXVIII. The œfophagus of neftlings more efficacious in this refpect   -   -   83
LXXIX. The whole length capable of digeftion   84
LXXX. The craw of gallinaceous birds does not digeft food   -   -   85
LXXXI, LXXXII, LXXXIII. Convenient mode of procuring gaftric fluid from crows without killing them. Its abundance. Its qualities. Is continually fecreted into the ftomach   -   86--88
LXXXIV. The œfophageal fluid procured in the fame way. Its fmall quantity. Bile gets into the ftomach. The reafon why the ftomach digefts fafter than the œfophagus   -   -   89
LXXXV. Gaftric fluid out of the body and in the cold, not more efficacious than water   -   91
LXXXVI. But when heat is applied, it then produces folution---Difference between its effects and thofe of water   -   -   93
LXXXVII. Speedy concoction of animal and vegetable matters by the gaftric fluid in the fun   -   94
LXXXVIII, LXXXIX, XC, XCI. Flefh immerfed in gaftric fluid, not diffolved in the fpace of a few hours, within tubes perfectly clofe and introduced into the ftomach. Some infufficient conjectural explanations of this phænomenon. The true reafon. Reflection on the importance of heat in thefe experiments   -   97--100

XCII.

XCII. Gastric fluid diluted with a great deal of water, produces solution in a brisk heat - 102
XCIII. Herons have an intermediate stomach. Description of it. Liquor secreted from the nervous coat into the cavity of the stomach, not by glands, but probably by arteries - - 103
XCIV. Stomach of herons always contains gastric fluid. Its qualities. Gall-bladder. The cystic duct probably inserted into the duodenum - 105
XCV, XCVI. Description of the œsophagus. Its follicles and liquor - - 106--107
XCVII. Stomach of herons compresses its contents. Digestion, however, does not depend on this action, but on the gastric fluid alone - 107
XCVIII. That of the heron more efficacious in dissolving bone than that of the crow - - 110
XCIX, C. The œsophagus of herons capable of producing a sensible degree of digestion - 111--113
CI. Proportion between the concoction of the œsophagus and stomach - - - 114
CII, CIII. Comparison between birds with muscular and those with intermediate stomachs, with respect to digestion - - 116--117

DISSERTATION III.

OF THE DIGESTION OF ANIMALS WITH MEMBRANOUS STOMACHS. THE FROG. NEWT. LAND AND WATER-SNAKE. VIPER. FISHES. SHEEP. OX. HORSE.

Reasons for treating this subject in several dissertations - - 119
CV. Singular way in which the gastric fluid of the frog in a day's time begins to dissolve flesh - 120
CVI. In a longer time it dissolves it completely without the action of the gastric muscles. Slowness of this process - - - 121
CVII. In time it dissolves bone - 122
CVIII. The gastric fluid of water-newts more speedy in producing its effects than that of frogs - 124

CIX

CIX. Difcovery of two fpecies of worms in the ftomach of this animal - - 126
CX. Defcription. Reafon for fuppofing that one fpecies is hermaphrodite and oviparous - 127
CXI. Stomach of this animal the refidence of thefe worms - - - 129
CXII. Similar worms between the internal and nervous coat in crows - - 130
CXIII. This is a certain proof, that the ftomach of the water-newt has no fenfible action - 131
CXIV. The reafon why infects that ferve the newt for food are digefted, and yet this never happens to the worms - - - 133
CXV, CXVI. Defcription of the ftomach and œfophagus in fome land-fnakes - 134--136
CXVII. Means contrived by the author, for obferving the various changes the food undergoes in the ftomach of ferpents without killing them - 137
CXVIII. Gaftric fluid of itfelf capable of digefting flefh in certain land-fnakes. Slownefs of this procefs 138
CXIX. The fame lefs flow, as the meat is lefs tough, and the gaftric fluid has freer accefs - 139
CXX. Œfophagus and ftomach of water-fnakes (denominated *natrices*) very like thofe of land-fnakes 140
CXXI. In them digeftion is the effect of the gaftric fluid alone - - ib.
CXXII. Probable arguments that this fluid diffolves bone alfo - - 142
CXXIII. Analogy between the gaftric fluid of this and other animals - - 143
CXXIV. Vipers refemble fnakes in the form of the œfophagus and ftomach, and in the mode of digeftion 144
CXXV. No digeftion in the œfophagus of thefe animals - - 145
CXXVI. Digeftion quicker in the warmer feafons 146
CXXVII, CXXVIII. Inftances of flefh lying a long time in the ftomach of thefe animals without putrifying - - - 148--149
CXXIX. Of digeftion in the eel - 149
CXXX, CXXXI. Defcription of the ftomach and œfophagus of the carp. The fource of the gaftric fluid - - 150--153

CXXXII.

cxxxii, cxxxiii. Defcription of thefe parts in the barbel and pike - - 154
cxxxiv. Digeftion in fifhes the effect of the gaftric fluid. Origin and progrefs in a pike - 155
cxxxv. The fame in a carp. The inferior part of the ftomach digefts more rapidly than the fuperior. Some degree of digeftion in the œfophagus. Proof that in fifhes, ferpents, the newt, and the frog, digeftion is independent of trituration - 156
cxxxvi. Two experiments of Reaumur on fheep 157
cxxxvii. Reaumur's experiments repeated fuccefsfully - - 159
cxxxviii. Doubts whether they are decifive in favour of trituration - - 161
cxxxix. Important circumftance overlooked by the French naturalift, which proves digeftion in fheep to be folely owing to the gaftric fluid - 162
cxl. Confequences of thefe experiments - 166
cxli. The gaftric fluid of fheep diffolves other fubftances befides herbs - 168
cxlii. An incipient digeftion obtained out of the body. Heat neceffary for this - 169
cxliii. The gaftric fluid is the caufe of digeftion in the ox and horfe - - 172
cxliv. Ruminating animals very much refemble birds with gizzards, with refpect to the action of the gaftric fluid - - 173

DISSERTATION IV.

THE SUBJECT OF DIGESTION IN ANIMALS WITH MEMBRANOUS STOMACHS CONTINUED. THE LITTLE OWL, THE SCREECH-OWL, THE FALCON, THE EAGLE.

cxlv. Recapitulation of Reaumur's experiments on the digeftion of animals with membranous ftomachs 175
cxlvi. Birds of prey. The gaftric fluid of the little owl incapable of digefting fome vegetable fubftances 178
cxlvii. Though capable of producing this effect on bone. The ftomach has no triturating power 180

CXLVIII.

CXLVIII. Contrivance of the author for bringing up tubes out of the ſtomach of birds of prey at pleaſure. Gradual ſolution of bone and fleſh in tubes by this owl - - ' 180
CXLIX. Inexhauſtible ſource of the gaſtric fluid---Properties - - 183
CL. It diſſolves fleſh out of the body - 185
CLI. Deſcription of the œſophagus and ſtomach. Source of the gaſtric fluid - 187
CLII. Morbid condition of a ſcreech-owl, that rendered the gaſtric fluid incapable of digeſting fleſh 189
CLIII. Which is very efficacious in health - 190
CLIV. Then even bone is readily diſſolved. The œſophagus, in one ſpecies of ſcreech-owl, diſſolves fleſh nearly as well as the ſtomach - 191
CLV. Artificial digeſtion with the gaſtric fluid of this ſpecies - - - 192
CLVI. Another ſpecies of ſcreech-owl, exactly like the preceding - - ib.
CLVII. Way to give tubes to a large falcon without-irritating it - - 194
CLVIII. Singular digeſtion of bone in tubes 195
CLIX. Of the ſame looſe in the ſtomach. Hard bones long in being diſſolved - 198
CLX. Soft ones the contrary - 199
CLXI. Enamel of the teeth not diſſolved - 200
CLXII. The ſame thing with reſpect to horn, and the cartilaginous coat of the gizzard. Tendon digeſted 201
CLXIII. Leather not digeſted. Another kind digeſted 202
CLXIV. Gaſtric fluid of the falcon does not digeſt vegetables - - 203
CLXV. Fleſh and bone digeſted out of the body in a ſufficient heat - - 204
CLXVI. Mode of digeſtion within and without the body alike. The craw does not diſſolve fleſh 205
CLXVII. Œſophagus and craw full of glands. Part of the gaſtric fluid comes from the ſtomach ib.
CLXVIII. Eagle - - 207
CLXIX. Its food. Courage in attacking and deſtroying animals larger than itſelf - 208
CLXX. Liquor running from the noſtrils into the mouth while it takes food. Conjecture on its uſe 210

INDEX.

CLXXI. Falsehood of the opinion, that birds of prey, and especially eagles, never drink - 211

CLXXII. Whether the eagle can live on bread. Its aversion for this food - - ib.

CLXXIII. When introduced into the stomach it is easily digested - - 212

CLXXIV. This is the mere effect of the gastric fluid alone. Manner of its action - 214

CLXXV. Gastric fluid of the eagle readily digests other substances besides animal ones. Some carnivorous birds turn frugivorous, and reciprocally 215

CLXXVI. Stomach of the eagle has some motion, but is incapable of triturating - 218

CLXXVII. Craw has no part in producing digestion 219

CLXXVIII. The manner in which flesh is decomposed in the stomach of the eagle - 221

CLXXIX. It acts in the same manner on flesh inclosed in tubes - - 223

CLXXX. Flesh digested in proportion as the access of the gastric fluid is more or less free - 224

CLXXXI. Readiness with which the gastric fluid insinuates itself into compact bodies - ib.

CLXXXII. This fluid soon dissolves hard bones. Singular phænomena attending these solutions 226

CLXXXIII. Gastric fluid of the eagle sooner dissolves bone than that of other birds---does not act on the enamel of the teeth - - 228

CLXXXIV. Whether it dissolves flesh also sooner. Mistake that may arise in this enquiry - 229

CLXXXV. A quantity of gastric fluid vomited spontaneously every day by the eagle. Its qualities 232

CLXXXVI. Artificial digestion. The gastric liquor does not easily freeze - - 233

CLXXXVII. Intestines, pancreas, and gall-bladder described - 235

CLXXXVIII. Small size of the stomach compared with the craw. Coats of the stomachs. Glands 237

CLXXXIX. Gastric juices made bitter by the bile. Liquor that oozes out from the inside of the craw. Œsophagus and craw without glands. Different liquors composing the gastric fluid - 240

DISSER-

DISSERTATION V.

THE SUBJECT OF DIGESTION IN ANIMALS WITH MEMBRA-
NOUS STOMACHS CONCLUDED. THE CAT. DOG. MAN.
WHETHER DIGESTION CONTINUES AFTER DEATH.

cxc. The gaſtric fluid of the cat the efficacious cauſe of digeſtion - - 243
cxci. Enquiry concerning the origin of this fluid 245
cxcii. Slight analyſis of the gaſtric fluid of the dog. It diſſolves fleſh, bread, and cartilage, incloſed in tubes ib.
cxciii, cxciv, cxcv. Boerhaave thinks that dogs cannot digeſt inteſtine fleſh and ligament. Is miſtaken. Cauſe of his error - 247--252
cxcvi. Undetermined queſtion, whether dogs can diſſolve bone - - 256
cxcvii, cxcviii. Experimental enquiry. Determination in the affirmative. Gaſtric fluid of ſome dogs corrodes the enamel of the teeth. At the time it diſſolves bone, leaves linen untouched - 258, 259
cxcix. Slight motion in the ſtomach during digeſtion 261
cc. Viſible, however, on opening the abdomen - 262
cci. The ſame in the cat. Incipient digeſtion produced by the gaſtric fluid out of the body - 264
ccii. Enquiry concerning the origin of this fluid 265
cciii. The chief of theſe experiments repeated upon Man. Neceſſity for this - 267
cciv. Maſticated bread incloſed in bags, perfectly digeſted in the author's ſtomach. Not completely, when the folds of linen are very numerous - 268
ccv. The ſame with reſpect to different kinds of fleſh boiled and chewed, and incloſed in ſingle bags 269
ccvi. The ſame in boiled fleſh not chewed 270
ccvii. As alſo in raw fleſh - 271
ccviii, ccix. Fleſh incloſed in tubes, digeſted in the author's ſtomach. This is the effect of the gaſtric fluid alone. Proofs that the human ſtomach does not triturate food - - 271--273
ccx. Confirmation of theſe proofs. Explanation of a ſingular phænomenon - 274
ccxi. Chewed fleſh and bread ſooner digeſts than that which is not chewed. Reaſon of this difference 275

ccxii, ccxiii. Flesh, membrane, tendon, cartilage, perfectly digested in the human stomach 277, 278,
ccxiv. Also tender bones, but not hard ones. The intestinal fluid has some part in producing these effects 278
ccxv. The author's method to procure the gastric fluid in a state of purity - 280
ccxvi. Its qualities. Incipient digestion out of the body - - 282
ccxvii. Confirmation of this experiment. Proof of the necessity of a certain degree of heat. Experiment proving, that a great degree of digestion is produced before the food passes to the intestines - 285
ccxviii. Recapitulation - 286
ccxix. Boerhaave's opinion concerning digestion 288
ccxx, ccxxi, ccxxii. Facts that oblige the author to relinquish this opinion. Refutation of an opinion, which confines the action of the stomach to the extraction of the juices of animal and vegetable substances - - 290--296
ccxxiii. Whether, according to Hunter, the great curvature of the stomach is dissolved after death; whence he infers, that digestion continues after death - - 298
ccxxiv. The author's observations do not exactly coincide with Hunter's - 299
ccxxv. Means to determine, whether digestion does really take place after death. Employed in a crow, and seem to prove the affirmative. Comparison between digestion in a dead and living animal 300
ccxxvi. No digestion in the œsophagus after death 302
ccxxvii. The influence of heat in these experiments. Digestion goes on equally well after death, whether the animal is killed immediately after having swallowed food, or food is introduced after the animal is killed 303
ccxxviii. Further experiments. When birds have digested the food to a certain degree, that process advances no farther, though it should continue longer in the stomach - - 305
ccxxix, ccxxx. Digestion after death in fishes and quadrupeds. Proofs of the necessity of heat to digestion in many animals - ib. 306
ccxxxi. Digestion after death does not go on so well when

when the ſtomach is taken out of the body. Reaſon why the ſtomach is not ſoon ſubject to be diſſolved as the food - - 307

DISSERTATION VI.

WHETHER THE FOOD FERMENTS IN THE STOMACH.

CCXXXII. Boerhaave thinks, that an incipient fermentation only can take place in the ſtomach 310
CCXXXIII. Different opinions of Pringle and Macbride. Their proofs, that digeſtion is a fermentative procefs deduced from food ſet in veſſels. Application to the human body - - 311
CCXXXIV. This procefs takes place in veſſels, whether water or ſaliva is employed - 314
CCXXXV. Doubts whether this would happen with the gaſtric fluid - - 316
CCXXXVI. Experiments that prove the negative 317
CCXXXVII, CCXXXVIII. Examination of the food during the time of digeſtion in ſeveral animals. No fermentation obſerved. Reaſons for doubting, whether even an incipient fermentation takes place 319--321
CCXXXIX. Whether any acid principle accompanies digeſtion. Proofs adduced by ſome in favour of this opinion - - 322
CCXL, CCXLI, CCXLII. This principle is very far from being obſerved in all food, and all animals. When it is obſerved, it diſappears at the completion of digeſtion - - 323--326
CCXLIII. This acid does not come from the gaſtric fluid, but the food - - 327
CCXLIV. Chemical analyſis, which ſhews, that the gaſtric fluid is neither acid nor alkaline, but neutral ib.
CCXLV. Argument of ſome phyſicians, in favour of a latent acid in the gaſtric fluid, deduced from the coagulation of milk in the ſtomach. Experiments with the internal coat of the ſtomach - 332
CCXLVI. The other coats do not curdle milk 334

CCXLVII.

CCXLVII. It is probable that this property is communicated to the internal coat by the gaftric fluid. This fluid curdles milk as well as rennet - 335
CCXLVIII. It is very doubtful, whether this property is a proof of latent acidity - ib.
CCXLIX. Facts adduced to prove, that digeftion is accompanied with putrefaction - 337
CCL. Digeftion is over in fome animals, before putrefaction can begin - - 339
CCLI, CCLII, CCLIII. Examination of feveral animals during the time of digeftion. No token of putrefaction - - 341--343
CCLIV. Except in fick animals. The facts mentioned in CCXLIX. examined and explained - 345
CCLV. The gaftric fluid is not only a menftruum, but antifeptic - - 347
CCLVI. It corrects putrefaction in phials 349
CCLVII. Putrefaction begins in the craw of gallinaceous birds, but is checked when the food paffes into the gizzard - - 350
CCLVIII, CCLIX. The ftomach has the power of correcting putrefied food - 351--353
CCLX. Reflection on the animals that feed on putrid flefh. Some animals that naturally abhor it, may be brought to feed on it - - 355
CCLXI. The antifeptic power of the gaftric fluid not owing to the falt it contains - 357
CCLXII. Error of a learned French writer, who fuppofes that a little common falt promotes digeftion 359
CCLXIII. The antifeptic property of the gaftric fluid cannot be explained by the fpecious theory of Macbride. The caufe unknown to the author 360
CCLXIV. Recapitulation - 362
Mr. Hunter's Paper - - 365
Dr. Stevens's Experiments - 375

END of the FIRST VOLUME.

www.ingramcontent.com/pod-product-compliance
Lightning Source LLC
Chambersburg PA
CBHW022118290426
44112CB00008B/714